HIGH SCHOOL LIBRARY
PORTAGE AREA SCHOOLS
PORTAGE, PA.

PROPERTY OF THE
COMMONWEALTH OF PENNSYLVANIA
ESEA TITLE IV-B 78-79

No. 799
$8.95

CB RADIO OPERATOR'S GUIDE 2ND EDITION

BY
ROBERT M. BROWN
&
PAUL L. DORWEILER

TAB BOOKS
Blue Ridge Summit, Pa. 17214

SECOND EDITION

FIRST PRINTING—OCTOBER 1975
SECOND PRINTING—JANUARY 1976
THIRD PRINTING—MARCH 1976
FOURTH PRINTING—FEBRUARY 1977

Copyright © 1969, 1975 by TAB BOOKS

Printed in the United States
of America

Reproduction or publication of the content in any manner, without express permission of the publisher, is prohibited. No liability is assumed with respect to the use of the information herein.

Hardbound Edition: International Standard Book No. 0-8306-5799-1

Paperbound Edition: International Standard Book No. 0-8306-4799-6

Library of Congress Card Number: 75-29687

Preface

With 40 CB channels as of January 1, 1977, Citizens Band radio has become the largest and most populous communications medium since the advent of the telephone. At latest count there were well over four million licensed CBers with over 15,000,000 transceivers in operation—and all this happened since 1974 when the 27 MHz band was "discovered" by millions of truck and auto drivers.

Why this fantastic acceptance? CB offers every citizen a short-range (five to 25 miles), low-cost (the average transceiver costs around $100), means of two-way radio communications for personal or business use on 40 CB channels. In just a few short years, CB has far surpassed all expectations in operational applications, growth, and technical advancements.

Equipment available to the average CB newcomer increases in sophistication and performance each year, with modern units offering anything from single-channel operation upwards to 23 or 40 channels. Yet, simultaneously, prices are tumbling to all-time lows.

There are no FCC examinations to take. No prior background in electronics is necessary. And already all major Detroit auto manufacturers are well on their way to offering in-dash, factory installed CB sets in their autos. Programs have been launched linking CB with law-enforcement agencies and service garages for motorists stranded on remote stretches of highway. But to go on would spoil the book for you.

This *CB Radio Operator's Guide* includes the newly revised FCC rules that went into effect on January 1, 1977, and they are marked by vertically rules lines in the FCC Rules and Regulations, Part 95, starting on page 193. Earlier changes affecting CB operation in 1976 and late 1975 are indicated in boldface (dark) type.

It is my hope that this guide will be of use to the beginning CBer and will help him get on the air and obtain the maximum of pleasure—with the minimum of "headache"—from his station equipment. Welcome to the realm of Citizens Band radio!

Robert M. Brown
KOD 2239

CONTENTS

1 Introduction To CB Radio
License Classes—Class D Operation—Operating Requirements—
Uses for CB Radio—CB and Public Service ... **7**

2 Getting Your License
License Renewal—Age Limit Controversy—How to Fill Out Your
License Application—Specific Instructions—What to Do Next—
Operation Under Another Person's License—FCC Engineer-In-
Charge Discusses CB Violations ... **34**

3 Selecting CB Equipment
Base Station/Mobile Transceivers—Hand-Held Walkie-Talkie
Transceivers—Buying Used CB Radio Gear—SSB Equipment ... **46**

4 Hand-Held Transceivers
What FCC Says About "No License" Transceivers—Advantages
of Part 15 Transceiver Operation—Available Types of Part 15
Units—Hand-Held Class D CB Transceivers—The Imports—
Enjoy Yourself ... **72**

5 All About CB Antennas
Beam Antennas—Coaxial Antennas—Grond Plane Antenna—
Vertical Antenna—Yagis and Quads—Stacked Yagis—Standing
Waves—What They Mean—Maximum Range—Antenna Limita-
tions—Radio Waves—Polarization—Citizens Band Antenna
Directory ... **94**

6 Coax Cable & Connectors
Using Military Specifications—Importance of the Dielectric—
Check Line Capacity—RF Attenuation—Conductors—Cable
Selection—Connectors ... **126**

7 The Installation
Mobile Installations—Mobile Antennas—Noise Suppression—
Base Station Installation ... **136**

8 Optimizing Your Station
Mobile System Improvements and Accessories—Power Monitors—Speech Compressor—Selective Calling—Periodic Checks: 12V DC Supply—Moisture, Dust, and Corrosion   **160**

9 Troubleshooting Your Rig
General Troubleshooting—Receiver Troubleshooting—Transmitter Troubleshooting   **167**

10 CB & Public Service: Providing Emergency Assistance
Your Emergency Monitoring Post—Your Mobile Communications Center—The Sign Marker—Let Everyone Into the Act   **175**

Appendix
Part 95 Citizens Radio Service   **193**

Index   **252**

Chapter 1

Introduction To CB Radio

The convenience of Citizens Band radio has made its personal applications important to the greatest number of present licensees. Practically everyone can use CB transceivers for the convenient short-range communication they offer. For example, a CB transceiver installed in an automobile is almost as convenient as a telephone. The hectic routine of Saturday morning errand-running is reduced to business-like efficiency as the housewife finds she can contact her harried husband by merely pressing a microphone button and talking. "The supermarket is out of Brand X detergent?...then get Sudzies instead." "Forget about picking Mary up after the movies, she's getting a ride with Doris' mother." "I just remembered that we are out of milk... better pick up two quarts on your way home."

The Federal Communications Commission has a rather formal definition for Citizens Band radio, but simply stated it means that almost any responsible citizen meeting a few basic requirements can obtain a license to operate a two-way radio. There are, of course, certain limits restricting the type of equipment, power ratings, and frequency. But for the most part you don't have to bother yourself with such things; Citizens Band radio is sufficient for the purpose for which it is intended as it comes from the box!

Until the advent of CB radio, only radio amateurs, police, taxicabs, and other public services were allowed to own and operate two-way radio equipment. The FCC realized that the general public also had a need to communicate and in 1947 they allocated a group of frequencies to be used for "Citizens Band

The now defunct Movin' On wasn't the only TV show to use CB. Stars of TV's "Flipper" series were careful to make certain that operations over CB—seen in many episodes—were legal to the letter of the law. This touch added an element of professionalism that many would-be CBers later employed in their own on-air operations. (Shown above: Brian Kelly and Luke Halpin, who also used CB in feature length movie of the same title.)

Radio Service." These frequencies were in the 460- to 470-MHz range, called UHF (Ultra-High Frequency).

Stations at that time were licensed only as Class A or B. The difference between them was the power limits they were allowed. Stations operating Class A were allowed a transmitter input power of 60 watts. Class B stations were limited to five watts input. However, there was no stampede to obtain a CB license at that time. The reasons were simple enough: the cost of equipment to operate at those frequencies was expensive and the usable operating range short. UHF transmission is classified as line-of-sight. In other words, you could transmit only from one point to another if there were no obstructions between them, such as the earth or buildings. Television signals operate on much the same principle, which is why you must have a high antenna with lots of gain to receive a TV station more than a few miles away. (A more detailed explanation of operating ranges and transmission is given in Chapter 5.)

In 1957 the FCC proposed use of the 11-meter amateur band

for CB operation—which would eliminate many of the problems associated with the higher UHF band. The 11-meter band includes the frequencies from 26.965 to 27.255 MHz. In 1959 the band was officially designated for use in the Citizens Band Radio Service. The rush was on. By the end of July that same year the FCC had processed more than 53,000 licenses. Licenses continued to pour in at an overwhelming rate and they now number well into the millions.

LICENSE CLASSES

Operation on Citizens Band frequencies today is limited to relatively short-range communication. This means you cannot legally talk to another station over 150 miles away. It is possible to hear "skip" stations from all over the country, or even from across the oceans, at certain times of the year and at certain hours. This is due to signal bounce (known as F2- and Sporadic-E reflection) from the upper regions of the ionosphere and from other atmospheric changes. It is more noticeable on some frequencies than it is at others, and—to

A CB transceiver installed in an automobile is almost as convenient as a telephone.

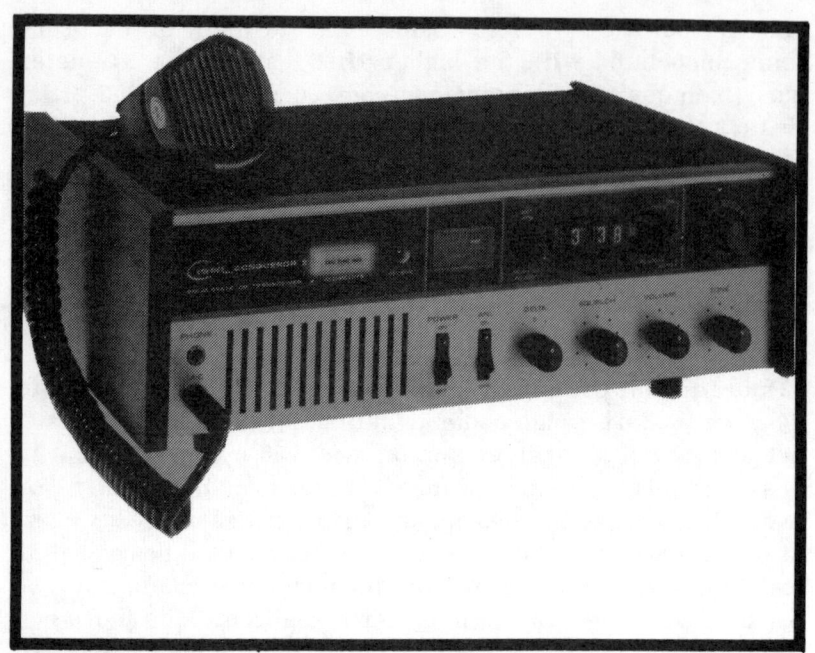

Courier's Conqueror II-deluxe 23 channel base/mobile CB rig with digital and illuminated ON-THE-AIR window.

confuse things even more—these "skip" signals change with the changes in the seasons. At times a foreign station may come in stronger than one only a few miles away. Regardless, it is illegal to communicate with "skip" stations.

CB licenses are classified into two types of service. These are: 1) fixed and 2) mobile. "Fixed service" is used to describe radio communications conducted between specified stationary points. "Mobile service" is radio communication between mobile and land stations or between two or more mobile stations. A base station is classified as a land station. All Class D stations (as well as Class B and C) are licensed as mobile stations only. However, they may be operated at fixed locations within the provisions listed in Part 95 of the FCC Rules and Regulations.

Besides classification of types of service, there are several classes of station licenses. There are: Class A, Class B, Class C, and Class D. Most all CB operators are licensed under Class D, but a simple brief explanation of each class should provide a better understanding of the overall structure of the Citizens Band radio service. Class A and B stations are licensed for UHF operation only, in the frequency range we mentioned earlier. However, as of March 18, 1968, applications for new Class B stations or modification of existing stations are not being accepted. Class C licenses are issued only for radio control of remotely operated devices such as model boats and aircraft. In addition, a spectrum of frequencies from 72 to 76 MHz has been delegated for remote control of model aircraft.

CLASS D OPERATION

The Class D license governing most CB owners and operators is designated for radio-telephone operation within the frequency band of 26.965 to 27.405 MHz at a transmitting power of 4 watts or less.* Class D stations licensed for radiotelephone operation means simply that the carrier—or channel frequency—is used to carry voice or audio tone only. The audio tones may be the type used to actuate receiver circuits to establish contact with another station such as is used on tone-signaling or selective-calling devices. But the FCC Rules state that the use of tones solely for the purpose of attracting attention or for remote control are forbidden under Class D operation. The transceivers used for Class D operation also must meet certain specifications and there are a few legal limits imposed on the transmitting antenna. Antenna limitations are discussed in detail in Chapter 5.

The CB transmitter is limited to no more than 4 watts output, modulation cannot exceed 100%, and the frequency tolerance of the transmitted carrier (channel frequency) must be held to within .005%. Normally, these tolerances are designed into the equipment when it is built. However, in some cases it is possible for the owner or operator to change these tolerances either through mis-adjustment, incorrect parts replacement, or a "homebrewed" circuit modification. Unless you have a First or Second Class Commercial Radiotelephone License, or work under the supervision of someone who does, it is illegal for you to change anything in the transmitter which could put it outside of the legal FCC limits.

OPERATING REQUIREMENTS

The FCC has set down a number of rules regulating opera-

*As measured at transmitter's antenna terminals.

CB transmitters can be adjusted (or repaired) only by service technicians holding FCC First or Second Class Commercial Radiotelephone licenses. Technician above is owner of specialty service center for CB troubleshooting: L.A. Communications, Los Angeles, California.

CB two-way radio enables pumping station manager to keep in contact with work crews shutting down water down water mains in Broadview, Illinois.

tion in the Citizens Band Radio Service, all of which have a definite purpose. Most of them are included to make your use of the available frequencies more efficient. The citizens radio service was initiated to provide the general public with a means of communication, but the FCC now allows its use as a hobby-type two-way radio system. You must call another

In small communities where public works crews have limited manpower, Citizens' Band radio provides an inexpensive way of keeping in touch with man performing routine duties. In case of emergency, this operator can be quickly summoned and directed to the location where he or his equipment is needed.

Public works repair team calls base for spare parts needed for repairs.

station using their designated callsign and identify your station with your own callsign. Communications between stations cannot exceed five consecutive minutes and there must be a 1-minute silent period before you can resume communications. The rules also state that you cannot use CB in violation of any federal, state, or local laws—which is pretty easy to understand. You cannot use it to stage a bank robbery, for poaching game, or for stealing cars. There also are rules against transmitting obscene language, broadcasting music, malicious interference, and a few other things, all of which are designed to provide you with the greatest amount of usable operation for your investment and needs. Remember, too, that on part of your application you sign a statement to the effect that you have read and understand Part 95 of the FCC Rules and Regulations. These rules also state that each licensee must maintain as part of his station records a current copy of Part 95. So violation of any of the restrictions cannot be blamed on ignorance of the rules. And by signing the application for a Citizens Band license, you indicate your responssibility for others operating equipment under your license regardless of whether they are employees or members of your immediate family.

USES FOR CB RADIO

There are numerous applications for CB radio, with new ones being found every day as people become more aware of the need to communicate. Users include sportsmen who install units in their boats, campers, cabins, and even in mobilized snow machines. CB radio is especially useful in delivery trucks, service garages, and repair shops to radio dispatch repairmen and tow trucks. Many construction companies and equipment installers use hand-held units for communicating with a central control station. Citizens Band radio is also used by hotel and motel operators for contacting their maintenance men and for making reservations with CB equip-

Thirteen feet below the surface, Broadview, Ill., these two public works crewman use a CB transceiver to communicate with workers on the ground several hundred yards away. The noise of compressors pumping fresh air into the sewer makes shouting impossible.

ped motorists. A housewife thinks about how she could keep in touch with her husband while he's on the road. A farmer thinks about all the time and money he could save if he could communicate with the farm house while he's out in the field. And a contractor considers the ways he could shuffle equipment and cut down the time costs—if only he could coordinate his operations with two-way radio.

Citizens Band radio isn't the only kind of two-way radio which these people could use, but it's certainly the least expensive to purchase, license, and operate. Because two-way radio is now within the reach of practically everyone, more and more people are taking advantage of CB's three attributes: control, convenience, and safety. For example, a small-town delivery service has no need for the privacy and 40-mile range offered by police-type two-way radio. And yet the owner of the service realizes that it would increase efficiency if he could call drivers making deliveries in the area of customers telephoning for additional deliveries or pickups. Without radio the drivers must make their delivery rounds, return to the garage for new orders, and then go out on the road again. This results in expensive backtracking. CB radio gives the business owner just the amount of communications assistance he needs—at about one quarter the cost of police-type two-way equipment. CB also meets the needs of service stations, salesmen, doctors, schools, construction firms, farmers, plumbers, and all the many other small and large businesses who need two-way radio but can't justify the thousands of dollars that police-type equipment requires.

Sailing craft have been particularly difficult to equip with two-way radio because they seldom carry marine batteries except when equipped with an emergency engine. For larger cabin craft, the new 5-watt equipment can be used with portable packs. Channel 13 has been the informal marine Citizens Band frequency. However, the advantage of Citizens Band radio is that since there are so many licensees, a small craft user need only switch channels in an emergency until he hears a station and call for help. In most sea coast areas, informal and formal monitoring services have been established for Channel 13.

Other recreational uses of Citizens Band radio can be found in hunting and fishing. In these sports it's not uncommon for groups to spread out, only to have one of its members encounter some kind of trouble out of earshot. Each winter the news media abound with stories of hunters dying of exposure, coronaries, and other accidents, where if they had been able to call for help they would be alive today. Here, the new high-power Class D portable units are excellent insurance. Their efficient transmitters have a range of several miles over even rough terrain. And they're reliable enough to risk your life on—which many hunters presently do with less reliable imported "toy" models.

From the commercial aspect, control is most important.

Each winter the news abounds with stories of children and sportsmen dying of exposure. Use of CB transceivers is a "must" for such expeditions. Even if a hunter becomes lost, he can be in immediate contact with his CB equipped buddies, who can then summon whatever assistance is required.

Pleasure craft are always likely places to find CB equipment in daily use not only from the "fun" standpoint, but also as a means of summoning aid should a water accident or sudden inclement weather develop.

Naturally, many major industries have some form of FM communications equipment. Others may have some form of field telephone. Few can afford to have FM units on every vehicle. However, with Citizens Band transceivers, they can.

Rugged Citizens Band transceivers have been needed in the heavy construction field so that equipment can be coordinated for maximum efficiency. If, for example, a power shovel breaks down, it can shut down waiting loaders, trucks, etc. The shovel operator must be able to contact his supervisor immediately for a less strategic power shovel to replace it. Without Citizens Band equipment, the operator of the shovel may have to waste time walking or being driven around the site to locate someone to make a decision. With CB equipment he merely picks up the microphone and calls. A single work stoppage ended in short order can pay for the initial investment of the transceiver.

Such an environment requires a rugged transceiver with foolproof controls, with the ability to take abuse from many operators. This is why industry has produced fully solid-state units built with steel cases and as few as only three controls.

Material-handling communications problems cover a wide range, from the direction of lift trucks in giant warehouses to the direction of stock personnel in small businesses. In each case, the problem is how to make the most of machinery every moment. Citizens Band radio offers excellent, low-cost communications for every industry, large or small. Because most materials and stock handling is done inside steel buildings, transceivers inside are shielded from communications outside the building. In many cases, Citizens Band equipment will work as well as the most expensive FM units. In the case of fork truck direction, many full 5-watt CB transceivers are inexpensive enough to equip each truck with one. In this manner, the warehouse manager can quickly select the proper size truck for the job at hand, rather than waiting for its operator to drive within shouting distance. Changes in priority can be instantaneously relayed to the farthest corners of the warehouse so that operators can be diverted from

CB is used extensively for materials handling in many manufacturing complexes.

On today's farm, low-cost CB equipment has become an invaluable tool.

original assignments to more urgent ones. Stockroom direction with Citizens Band radio can result in handling orders in half the time now needed with present intercom systems, as well as simplify inventory control.

In a CB-equipped store, when a customer places an order —either over the counter or over the phone—the salesman merely calls the stock manager on the intercom and relays the order. The manager then puts out a general call to the stock boys: "Who's in the area of aisle 2, row 1?"...on one channel of his transceiver. The one nearest replies on that channel, receives his order, and selects the stock. In this manner, the manager can quickly report back to the salesman if the article is not in stock and request an alternate order. Stock boys don't waste time backtracking, and, most important, they can pick up several orders on a single trip instead of just one.

Whenever a stock boy discovers that a certain product has

Remember the Platformate commercials for Shell? Squires-Sanders (no longer manufactured) CB radio equipment was used in each vehicle as a hedge against something going wrong, leaving a driver stranded on the hot salt flats miles away from humanity with an overheated automobile. CB also assisted in coordinating the production of the filmed commercial by Shell's advertising agency.

For the plumbing, air conditioning and other service industries, CB can mean rapid job-to-job movement during crucial times of the year when business must be turned away because the manpower to handle it can't be seasonally hired.

reached a specified minimum he can switch to another channel and inform the accounting department that it is time to reorder. This multiple-channel operation—one for stock manager to stock boys and another for stock boys to accounting office—results in highly efficient management of time as well as a system for instant inventory control. This same kind of control is possible with fleet operations, be they truck or car. Delivery and pickup vehicles can be rerouted according to phoned-in orders without having to return to base. A single dispatcher with a map and stick-pins can pick the nearest driver and direct him to the new destination. Also, the dispatcher can keep track of drivers who "goldbrick" on company time by instructing each one to report in and out.

Now, this is old hat to many firms, but others—especially the smaller ones—are still convinced they can't afford such communications systems. Reliable, easy-to-operate Citizens Band equipment puts this system into the price range of the fleet operator who needs it most: the small one with only a handful of trucks or cars. In hundreds of thousands of rural areas, Citizens Band can invade FM's domination of the taxi industry. A carefully placed base station can result in ranges of up to 20 miles or more—more than adequate for such applications. And the channels will be completely free almost all the time.

Salesmen of all kinds are realizing the advantages of Citizens Band radio communications equipment. For example, it enables an insurance firm to quickly dispatch the calls of present or potential policyholders to salesmen in the field. With a check-in, check-out arrangement, each salesman finds out if there are any new business prospects in his immediate area. This idea isn't original—it is just made practical by Citizens Band radio. The potential for this kind of system exists in almost every vehicle-oriented sales organization. For the plumbing, air conditioning, and other service industries it can mean rapid job-to-job movement during crucial times of the year when business must be turned away because the seasonal manpower to handle it can't be hired. For automobile service stations it can mean more efficient towing operations with fewer instances when an attendant says, "Can't say when our truck will be along... he's out on a call right now."

CB AND PUBLIC SERVICE

Project HELP: The HELP (Highway Emergency Locating Plan) program is rapidly becoming more and more evident as

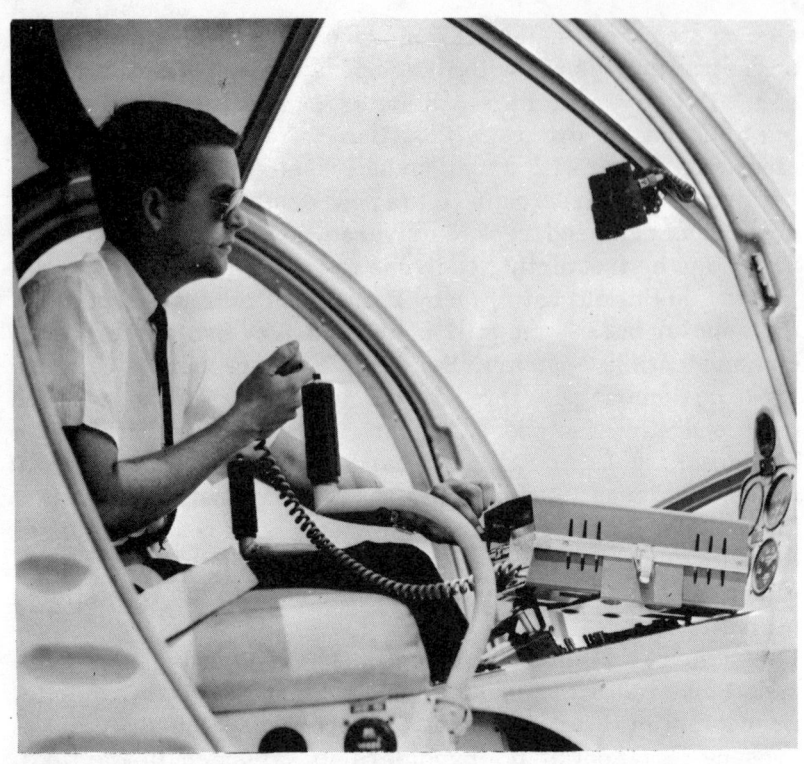

CB equipped helicopters are part of law enforcement's newest means of spotting crime in the streets in large metropolitan areas throughout the country. Criminal actions or suspicious acts are sighted from the air and radioed to headquarters where a squad car is immediately dispatched to investigate.

one travels the major highways across our country. It is a plan to provide assistance to motorists who become stranded, sick, or for some other reason need emergency aid while on the road. As it is now, a few police agencies, hospitals, garages, and volunteer citizens monitor the HELP channel in hundreds of cities and towns. Channel 9 has been selected as the HELP calling channel. You may have seen signs along the road indicating that a local radio club or service station is monitoring the HELP channel.

Women especially appreciate the safety afforded by Project HELP, originally developed by Detroit's Automobile Manufacturers Association. Even in areas where Project HELP is not yet in full swing, however, Channel 9 is being monitored by volunteer service-minded CB groups. Hand-held CB transceivers also offer an important degree of safety to women who must leave their homes at night, either to shop or to work. A 1.5-watt CB "walkie-talkie" weighing only about a pound has a range of three to five miles—more than adequate to summon aid in case of trouble, particularly if the walkie-talkie is tuned to Channel 9, commonly monitored in and around major cities. Channel 11 is the national calling channel.

The use of CB units for street protection is not new; police across the country have used this equipment for years. However, Citizens Band walkie-talkies are within the budget of the private citizen, whereas most police-type equipment is not.

Presently, only motorists with CB radio equipment can make

REACT, means instant aid for stranded motorists equipped with CB gear capable of operation on Channel 9. In many areas signs are posted, indicating that a given channel is under constant surveillance by local police and service garages.

25

use of this emergency aid, but because the idea itself has gained such interest, automobile manufacturers, auto clubs and CB radio manufacturers are working on ways to equip every automobile with some type of receiver and transmitter which can be used to call for help in time of need. Some companies have designed units which use the automobile's broadcast radio as a receiver and contain a small solid-state transmitter. These are considered as a converter/transmitter combination. The units designed for this type of operation are normally more limited in range than an average CB transceiver depending on location, terrain, and other operating conditions. The ideal situation for use of low-power units of this type would be to have monitoring stations at regularly spaced intervals along our major highways, but such a plan would undoubtedly be costly. Some manufacturers have been working on ways to overcome the problems concerned with making the HELP plan available to everyone. One of the solutions offered is to build fixed emergency calling stations periodically spaced along the highway where all a person has to do is walk up and push a button or flip a switch and an alarm would be set off in the nearest town. These might be similar to the fire call boxes we have all seen on street corners and in buildings. Whatever the answer, the HELP program is gaining popularity and it will need the cooperation of all CBers if it is to succeed.

REACT: While the non-profit REACT programs (for Radio Emergency Associated Citizens Teams) is not the most complex or sophisticated, it is easily the best example of what volunteer channel-monitoring can accomplish. Consisting of about 1400 24-hour stations ready to provide CB Channel 9 communications, augmented by a hard-core active membership of 40,000, REACT has handled over one million on-frequency requests for routing assistance, road service, and emergency aid (ambulance, fire, police, etc.).

Realistically, when viewed in terms of a contender for a countrywide motorist emergency service, however, its abilities pale in comparison with total needs. For true participation the REACT member must be a licensed CBer familiar with equipment limitations and have the conventional under-the-dash transceiver handy on all occasions. The fact remains that REACT is the only radio-aid system now in successful operation on a large scale in the United States.

Highway Emergency Radio (HART): Truck drivers travel many millions of miles each year on U.S. highways. With the concept that the addition of CB radio equipment could make each vehicle an effective emergency unit, several members of the trucking industry have activated a program known as HART—Highway Aid by Radio Truck. HART members are instructed in First Aid, carry First Aid equipment, and are schooled in emergency procedures. Additionally, the average truck driver has a knowledge of mechanics and can generally determine service required in automobile breakdowns.

Among the rules established for HART members: 1) proper

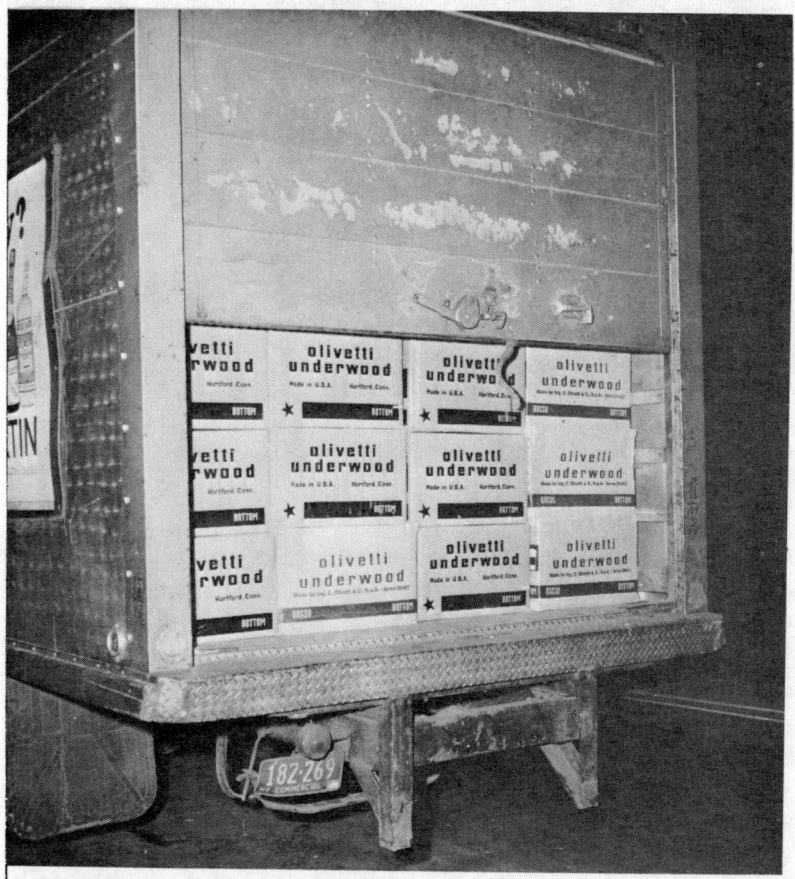

Photograph from New York City Police file showing a vehicle recovered as a result of radio-alert system (see "HART/Highway Aid by Radio Truck"). Truck contained thousands of dollars worth of Olivetti Underwood typewriters.

Frank Mack, controller of Empire State Highway Transportation Association, listens to driver James Carroll check on details of "alert" message.

CB licensing; 2) promotion of safe driving and exceptional care of member vehicles; 3) all possible courtesies and aid extended to those in need: 4) detailed handling of major emergencies, including the notification of authorities; 5) protection of exposed personal property of victims at an accident scene; and 6) extension of HART services without payment. CBers employed in the trucking industry can obtain more information by writing to HART, Box 141, Pontiac, Michigan.

The DAIR Program: Another CB plan called Driving Aid, Information and Routing (DAIR) is an experimental system resulting from more than ten years of work by General Motors Research Laboratories, Detroit. DAIR is designed to assist CB-equipped motorists in case of an emergency, in addition to providing automatic routing for extended trips and warning of approaching traffic-sign and speed limit changes. Motor-

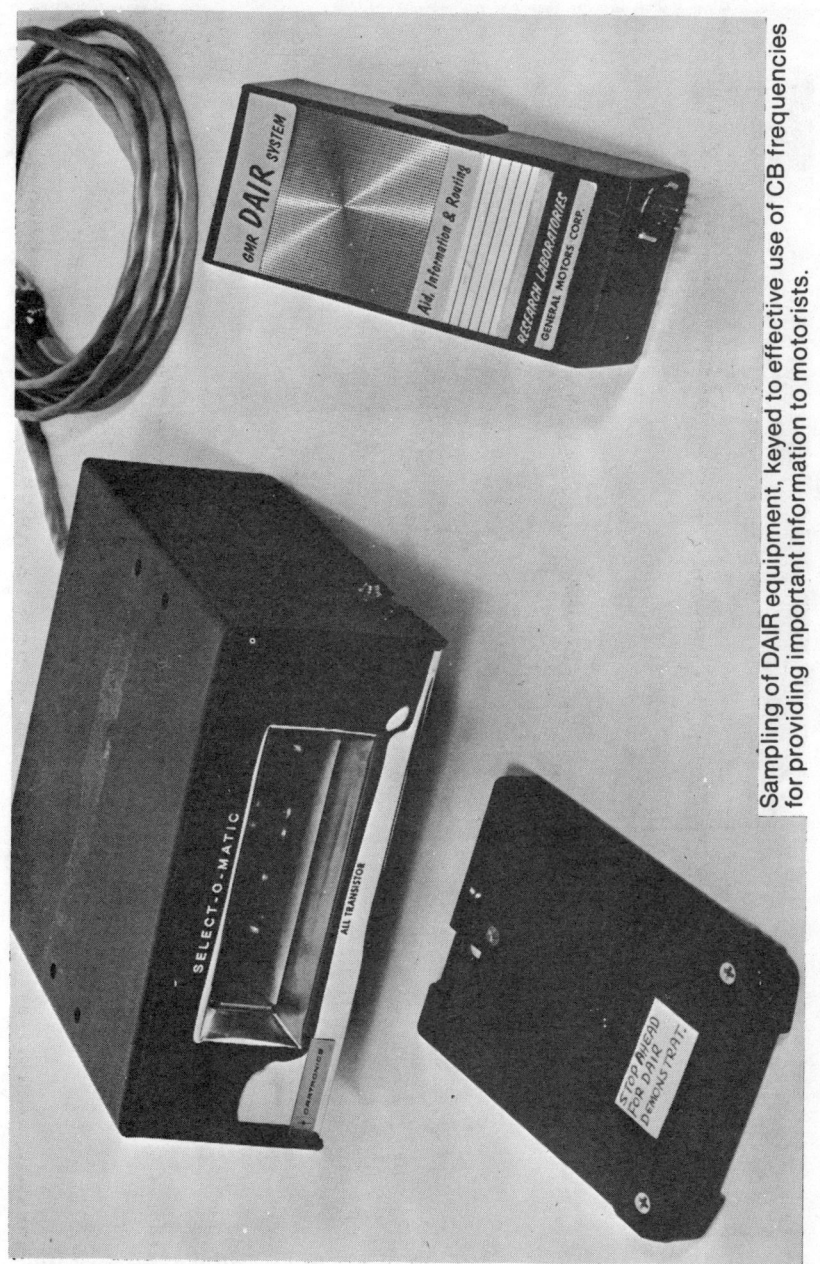

Sampling of DAIR equipment, keyed to effective use of CB frequencies for providing important information to motorists.

Best of all, CB is far from a man's-only communications medium!

Base station Citizens' transceivers are usually as small as a telephone and just as easy to operate. Here, Kenneth Cameron, superintendent of public works, Broadview, Ill. keeps in touch with work crews. With the addition of a nickel-cadmium battery pack, his transceiver may be used as a powerful hand-carried unit. With a good antenna, such 4-watt transceivers have a range of up to 20—50 miles. When installed in a vehicle, or carried as a portable, such units have ranges of about 10 miles, varying with terrain.

ists in the DAIR program would make use of a special console with a related visual sign-minder mounted on the dashboard. This console would be equipped with 1) a microphone for direct voice communications with an information center, 2) a telephone-type dial for transmission of coded messages, and 3) a punch-card reader for planning which route to take.

Community Radio Watch: The Community Radio Watch program, initiated and sponsored by Motorola Communications and Electronics, Inc. Chicago, Ill., encourages citizens, particularly those who use "any type of two-way radio," to support the police in maintaining law and order. Drivers of radio-equipped vehicles serve as "eyes and ears" for the police. Since Community Radio Watch was introduced in 1966, many cities and companies with personnel have started local programs.

Three plans have been suggested for establishing a Community Radio Watch on a local basis: that the program be handled 1) through the mayor's office, or 2) by a mayor's committee (sheriff's department, etc.), or 3) by a local organization such as the Junior Chamber of Commerce. The program is open to anyone who uses two-way radio—including licensed CBers interested in participating. For full details, write to Community Radio Watch, Motorola Communications and Electronics, Inc., 4501 W. Augusta Blvd., Chicago, Ill. 60651. In addition to these programs, several others exist which are well worth investigating after you obtain your CB license. Check the monthly CB magazines for a handle on what's being done in your locale.

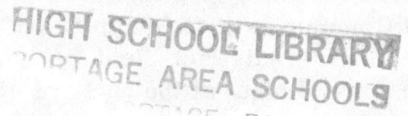

Chapter 2
Getting Your License

Almost any citizen of the United States over 18 years of age is eligible for a Citizens Band license, providing that person does not represent an alien or is a representative of a foreign government. There are other exceptions listed in Part 95.7 of the FCC Rules and Regulations calling out further general citizenship restrictions which apply to persons associated with foreign governments.

No one is permitted to hold more than one Class C, and one Class D station license. If one member of a family obtains a license, the other members of the family may operate the equipment under the same authorization. Applicants for Class C licenses need be only 12 years of age. The fee for all CB licenses is $4.00 and should be sent with the application to the FCC in the form of a check or money order (made payable to the "Federal Communications Commission"). Never send cash. Applications can be sent directly to the Federal Communications Commission in Gettysburg, PA or to your nearest district field office. The address of these offices are listed at the back of this book.

LICENSE RENEWAL

An applicant for a CB license need only renew the license once each five years from the date it was originally issued, renewed, or modified. It is no longer necessary to renew or apply for a modified license when you move. In the case of a Class D station, a change of address simply requires a letter giving your name, address, callsign and class of license listed as they appear on the FCC's records, along with the address for your new location.

If you decide to add more transmitters to your existing operation, you must submit a new application for modification. It is wise to think ahead when first applying for your license to include enough transmitters to cover any future needs which may arise during the term of the license.

AGE LIMIT CONTROVERSY

As indicated earlier in this chapter, the age limit for an authorized licensee is 18 years or older except for Class C, which is 12 years. There has been some controversy over the minimum age limit requirements for Class D because the members of a family can operate on a license held by another in the family. The question is: Why are people under 18 years judged unqualified when at the same time they can legally operate under a license held by a member of their immediate family?

The FCC rules state that the age requirement for a person wanting authorization for a Class D license is 18 years. It

Who is eligible for a CB license? People from all walks of life can and do apply for FCC "tickets" daily. Represented above are (from left) a repairman, nurse, fireman, farmer, construction worker, housewife, painter, deputy sheriff, teenage bird trainer, security guard, lift truck operator, and business executive.

does not say that you have to be 18 years old to operate a CB radio. If you hold the Class D license and other members of your family—regardless of their age—operate your equipment under your license, you are responsible! Remember, you indicated on your FCC license application that you accept full responsibility for the operation and control of any station licensed to you according to the FCC rules. Normally, persons under 18 years of age would be members of a family already licensed. Regardless of whether they are six of 60, if they operate under your license, you are responsible. Age has nothing to do with where your responsibility begins or ends.

HOW TO FILL OUT YOUR LICENSE APPLICATION

The standard application for a Class C or D station license is FCC Form 505. If you recently purchased a new CB transceiver, chances are that the form was packed with it. If not, you can obtain one by writing to your nearest district field office or checking with your local electronic dealer. CB dealers usually have them on hand as a service to prospective customers. Once filled out and in the mail, you can get on the air right away by filling out temporary permit 555B.

The application is not difficult to fill out, although the required information must be accurate. A worksheet is included with most Form 505s which you can keep for your records. The best procedure is to fill out the worksheet first and then transfer the information to the original Form 505. The application must be typed or clearly printed and should be filled out completely to avoid your getting it back, thus causing an unnecessary delay in obtaining your license. And remember to include the $4.00 fee!

Be sure to include all details of the station to be operated under your license. When the license is issued it will contain all your operating authority for a specific location or area and will supersede any previous authorization which you may hold of this class for the same radio system or group of transmitters. Read and answer all Form 505 questions carefully. Your signature certifies that the statements you have made are true. Any willful false statements made on this form are punishable by fine and imprisonment. U.S. Code, Title 18, Section 1001.

SPECIFIC INSTRUCTIONS

Items 1 and 3: If you are operating under a business name, give only that name in Item 1. If you are operating as a sole proprietorship or as a partnership with a business name, enter the name of the owner or partners in Item 3. If it is a partnership without a business name (see instructions for Item 4), give the name of one partner (as an individual) in Item 1 and enter the names of the other partners in Item 3. If applying only as an individual, give your last name, first name, and middle initial in the space provided. If you are a married woman, use your given name in the proper item, such as Doe, Mary S., not Doe, Mrs. John. If the applicant is a governmental entity, give the full legal name in Item 1 as stated in its authorization document; if a corporation or association, give the name in Item 1 exactly as it appears in the articles of incorporation or association.

Item 4: Give your permanent mailing address. If a P.O. or RFD# is used, fill out Items 8-10. Do not give an A.P.O. or

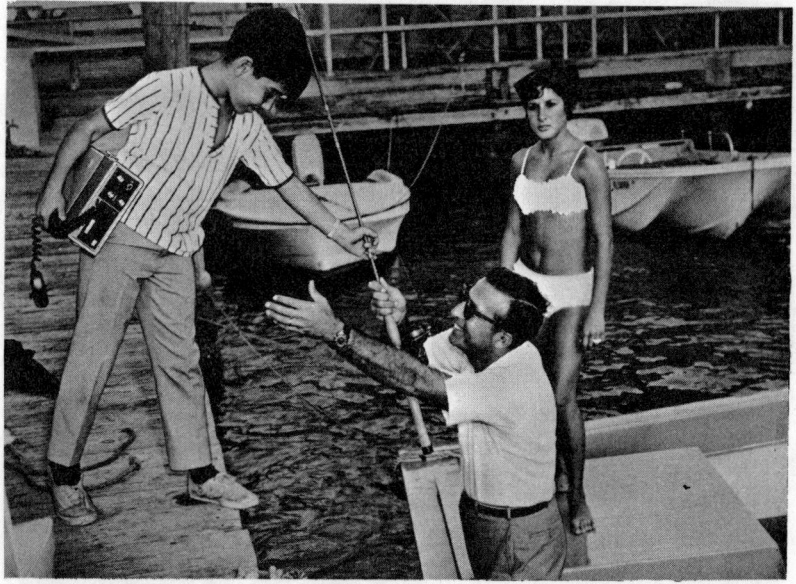

According to FCC regulations, all members of the family can operate Class D CB equipment, providing actual license holder (father above) is over 18 years of age and assumes responsibility for operations conducted.

United States of America
Federal Communications Commission

Form Approved
GAO No. B-180227(R01 02)

FCC FORM 505
December 1974

APPLICATION FOR CLASS C OR D STATION LICENSE IN THE CITIZENS RADIO SERVICE

Instructions

A. Use a typewriter or print clearly in capital letters. Stay within the boxes. Skip a box where a space would normally appear.
B. Sign and date application.
C. Enclose appropriate fee with application. DO NOT SUBMIT CASH. Make check or money order payable to Federal Communications Commission. No fee is required for an application filed by a governmental entity. For additional fee details, including amount and exemptions, see Subpart G of Part I, FCC Rules and Regulations.
D. Do not enclose order form or subscription fee for FCC Rules.

E. MAIL APPLICATION TO FEDERAL COMMUNICATIONS COMMISSION, GETTYSBURG, PA. 17325.

1. Complete if license is for an individual
Applicant's First Name   Init.   Last

2. Date of Birth
Month   Day   Year

3. Complete if license is for a business
Applicant's Name of Business, Organization, Or Partnership

4. Mailing Address (Number and Street) If P.O. Box or RFD# Is Used Also Fill Out Items 8—10

5. City   6. State   7. Zip Code

NOTE:
Do not operate until you have your own license Use of any call sign not your own is prohibited

8. If Item 4 is P.O. Box or RFD#, Give Address Or Location Of Principal Station

9. City   10. State

11. Type of Applicant (Check one)
☐ Individual   ☐ Association   ☐ Corporation
☐ Business Partnership   ☐ Governmental Entity
☐ Sole Proprietor or Individual/Doing Business As
☐ Other (Specify) _____

12. This application is for
☐ New License
☐ Renewal
☐ Increase in Number of Transmitters

IMPORTANT
Give Current Call Sign

13. This application is for (Check only one)
☐ Class C Station License (NON-VOICE—REMOTE CONTROL OF MODELS)
☐ Class D Station License (VOICE)

14. Indicate number of transmitters applicant will operate during the five year license period (Check one)
☐ 1 to 5   ☐ 6 to 15   ☐ 16 or more (Specify No. ____ and attach statement justifying need.)

15. **Certification**   I certify that:
• The applicant is not a foreign government or a representative thereof
• The applicant has (or has ordered from the Government Printing Office) a current copy of Part 95 of the Commission's rules governing the Citizens Radio Service
• The applicant will operate his transmitter in full compliance with the applicable law and current rules of the FCC and that his station will not be used for any purpose contrary to Federal, State, or local law or with greater power than authorized.
• The applicant waives any claim against the regulatory power of the United States relative to the use of a particular frequency or the use of the medium of transmission of radio waves because of any such previous use, whether licensed or unlicensed

WILLFUL FALSE STATEMENTS MADE ON THIS FORM OR ATTACHMENTS ARE PUNISHABLE BY FINE AND IMPRISONMENT. U.S. CODE, TITLE 18, SECTION 1001.

16 _____
Signature of individual applicant or authorized person on behalf of a governmental entity or partnership or an officer of a corporation or association.

17 Date _____

ORDER FORM   Please Print Or Type

Please enter _____ subscription(s) to Volume VI, containing Parts 95, 97 and 99 of the Federal Communications Commission Rules and Regulations ($5.35 per domestic subscription which includes U.S. Territories and for Canada and Mexico. $6.70 per other foreign subscription.)

Name   First, Last

Company Name Or Additional Address Line

Street Address

City   State   Zip Code

☐ Remittance Enclosed (Make checks payable to Superintendent of Documents)
☐ Charge to my Deposit Account No. _____

MAIL ORDER FORM TO:
Superintendent of Documents
Government Printing Office
Washington, D.C. 20402

Navy number or an overseas address. If a P.O. Box or RFD# is used, refer to Items 8-10. Such addresses will otherwise result in rejection of the application. If the license is to be mailed to such an address, so indicate under remarks.

Item 11: Check only one box. If you as the applicant represent:

- (a) An individual applying for personal or for business use without a trade name, check "Individual."
- (b) A legal business partnership, check "Business Partnership." A station licensed to a partnership may be used only for the business of that partnership and may not be used in connection with the personal activities of the partners.
- (c) An individual applying under a trade or business name, check "Individual/Doing Business As."
- (d) An unincorporated association, check "Association."
- (e) An incorporated association, check "Corporation."
- (f) A city, township, or other governmental body, check "Governmental Entity."
- (g) A trust or joint venture, check "Other," specify classification in space provided and explain.

Item 12: The license for a station identified by callsign in this Item will be superseded by the new license and callsign. You cannot apply for an additional station license of a class you already hold. Instead, you may apply for modification of your present license(s), if any, to include all transmitters you expect to operate under each class during a 5-year term. Class B, C, and D stations may be operated anywhere in the United States.

Item 13: Check one box to indicate the class of station for which you are applying. A separate application is required for each class of station, as follows:

- Class C: Control operations only, in the 27-MHz range.

- Class D: Voice operations only, in the 27-MHz range.

- Applicants for Citizens Class A Station Licenses must use FCC Form 400.

Item 14: Give the total number of transmitters expected to be in operation at any one time during the next five years under this license. Attach a list showing proposed use if the application is for seven or more units on an individual basis (13 or more if applicant is a business, in which case describe).

If your are applying for a Class B or a Class C station license using frequencies in the 72-to-76-MHz band, the transmitting equipment you propose to use must either be on the Commission's "Radio Equipment List, Part C" or you must attach a complete description of your transmitter in accordance with Subpart C of Part 95 of the Rules. If you are applying for a Class C station license using frequencies in the 26-to-27-MHz band or a Class D station license, the transmitting equipment must either be crystal-controlled or must appear on the above-mentioned list. If not, you must attach a complete description as indicated for Class B above.

Item 15: This item is your certification as to whether or not the use of the station for which authorization is requested will conform at all times with the permissible communications set forth in the Commission's rules, including the prohibitions against nonsubstantive and "skip" communications. The 150-mile limit on communications has been set for convenience but reflects the requirement that communications be directed to stations within the direct groundwave coverage range. Goundwave communications of 150 miles are far beyond the normal capability of citizens' radio stations, and "skip" or "skywave" communications may not be expected to occur except at distances considerably greater than that.

Item 16 and Item 17: Self-explanatory.

WHAT TO DO NEXT

After you have Form 505 completely filled out, signed, and double-checked for omissions or errors, mail it to your nearest district field office or to the Federal Communications Commission, Gettysburg, Pennsylvania 17325. Form 505 is used also for renewals and modifications of Class C and D licenses and sent to the same address. License renewals should be submitted at least 60 days before expiration. It generally

CR-123B base station for SSB/AM operation includes plug-in microphone, crystal filters, S/RF meter and plug-in PA. (Courtesy Regency Electronics).

Where CB use will be in connection with a business such as shown above (Pizza Express' delivery service in Boston, Massachusetts), the FCC application must clearly indicate the company or corporate name. See "specific instructions" for details.

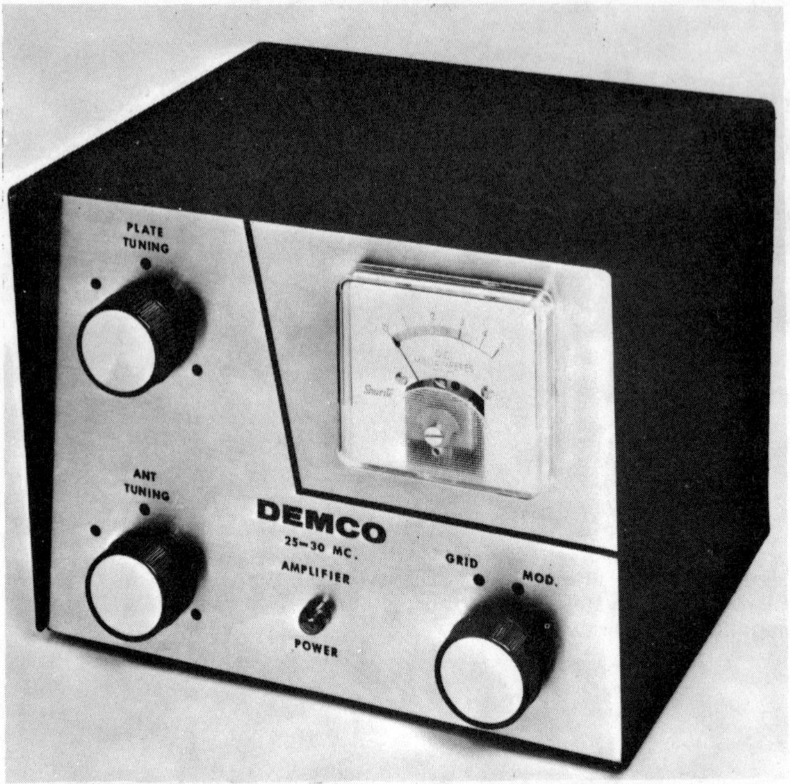

You can't fool FCC engineers by trying to hide an illegal device, such as the above linear amplifier (capable of multiplying the signal strength of a transmitter beyond that permitted under U.S. FCC law). Violators are subject to stiff fines, license suspensions and, sometimes, even jail terms. (Unit shown above is designed for legitimate use with 30-50 MHz Business Radio Service transceivers and 10-meter ham radio equipment; for those services it is legal; for CB it definitely is out.)

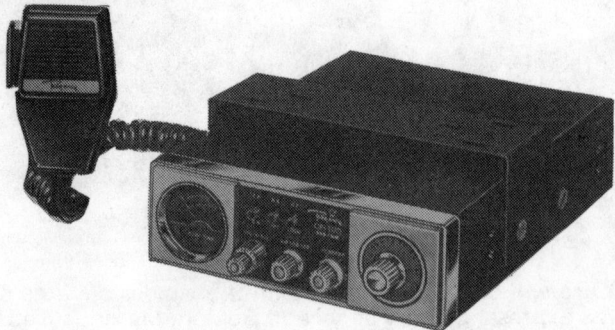

CR-123 SSB/AM mobile transceiver with S/RF meter and auomatic modulation control on AM. (Courtesy Regency Electronics)

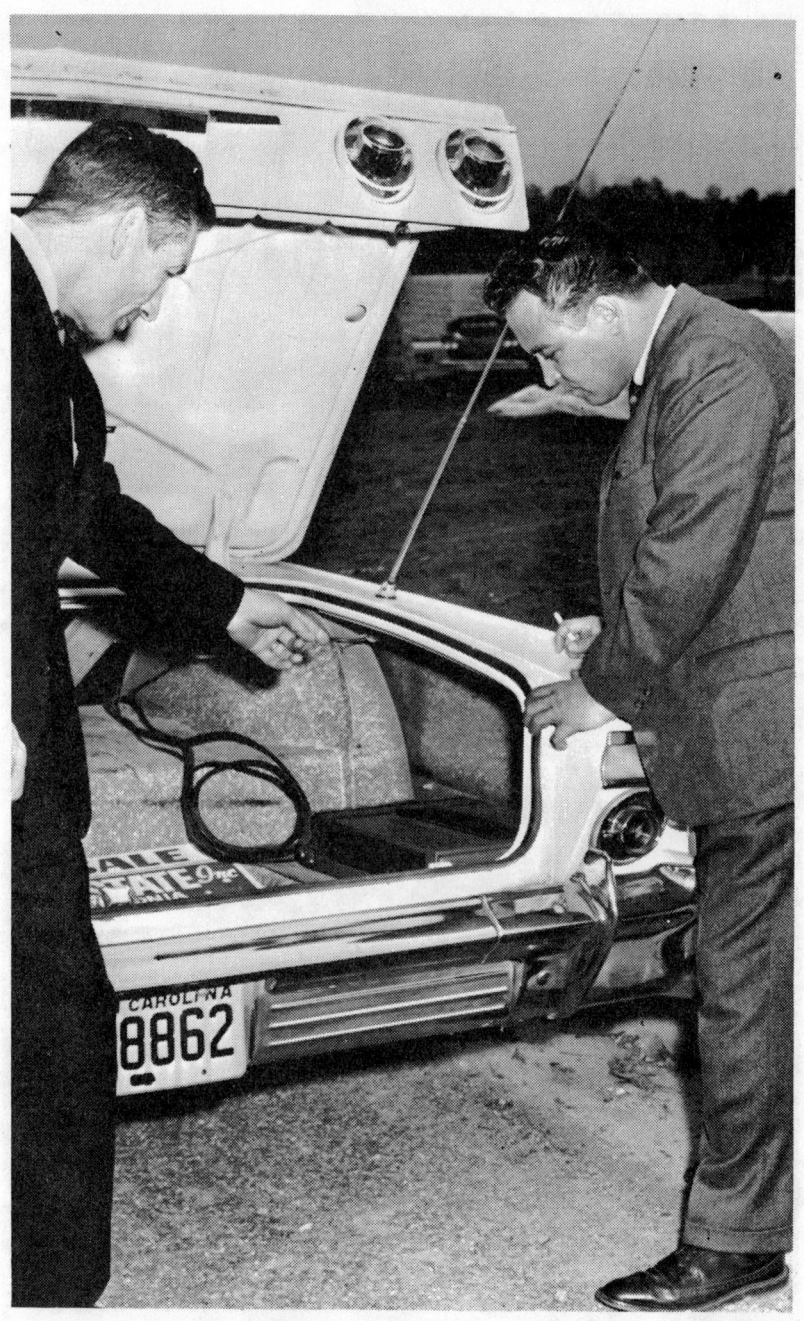
In photo above, FCC inspector catches CBer with an illegal 100-watt linear amplifier neatly tucked away in the trunk of his car.

Under Part 15 of the FCC rules and regulations, no license is required for low-power hand-held transceivers and no age limitation is imposed. Here, brother and sister communicate over quarter-mile distance without the necessity of using call signs. Rules already in effect will move all license-free walkie talkies to the new 49 MHz band by the early 1980's.

takes about four or five weeks to receive your license authorization back from the FCC. You cannot legally operate your transmitter without this authorization, but you can use the receiver to monitor the CB channels. When your license arrives, it must be posted where it can easily be seen at the normal fixed operating station. Licenses for equipment used as mobile stations do not need to be posted, but it must be kept as a permanent part of the station records.

OPERATION UNDER ANOTHER PERSON'S LICENSE

You've probably heard of people who bought CB radio equipment and operated it using another individual's license authorization (and call letters) until his own arrived. There are some pretty definite regulations in Part 95.87 concerning this and they specifically place responsibility on the person holding the license. In brief, the FCC amended the rules in September 1975, with the new requirement that each operator must use his own call sign.

There are situations where you may legally obtain permission to allow another person to operate under your license. But it requires a letter to the FCC, spelling out your reasons, how you propose to maintain control over the transmitter, and why that person must use your license instead of simply obtaining one of his own. Even if your reasons are valid and you do receive permission, it can still be revoked at any time by the FCC. You no longer can legally allow members of your own family to operate the equipment under your license. If your license is for business, your employees may operate the equipment—for business use. Part 95.87 establishes additional circumstances in which partners in a business, members of unincorporated associations and others may be authorized to operate under the control of a licensee with certain provisions. However, in all cases, the licensee is responsible for proper operation of all units of the station.

YOUR OPERATION

The September 1975 rule changes were the most sweeping ever in the annals of the citizens radio service. They provided for hobby-type use of Class D CB channels.

Chapter 3
Selecting CB Equipment

Selecting equipment at a well-stocked communications equipment dealer's store can be a confusing job because he may have Citizens Band transceivers from as many as 40 manufacturers. And since the differences which separate them are quite technical, you need to understand a little about CB equipment before making your choice.

Basically, there are three types of Citizens Band transceivers—1) license-free hand-held units rated at one-tenth of a watt power, 2) hand-held units more powerful than one-tenth of a watt which require an FCC license, and 3) mobile, base station, or portable 4-watt transceivers, also requiring an FCC license.

If your communications are going to take place over a range of a mile or less, the one-tenth watt units may be all you need. However, don't confuse these transceivers with the toy types often sold for $10 or less—these are sensitive, efficient communications tools with a much more dependable range than the dime-store units. Although they are more expensive, you may find it worth your investment to select transceivers powered by nickel-cadmium batteries rather than the throwaway kind used in portable radios. Nickel-cadmium cells can be recharged when they run down, saving you many times their original cost in the long run.

Should you need to talk over a distance of one and a half to three miles, the more powerful hand-held transceivers may be right for you. These often look much like the one-tenth watt types and cost only a little more, but they must be licensed with the Federal Communications Commission. If you need reliable communications over a 10- to 20-mile range, the 5-watt transceiver is the only type to choose. Different

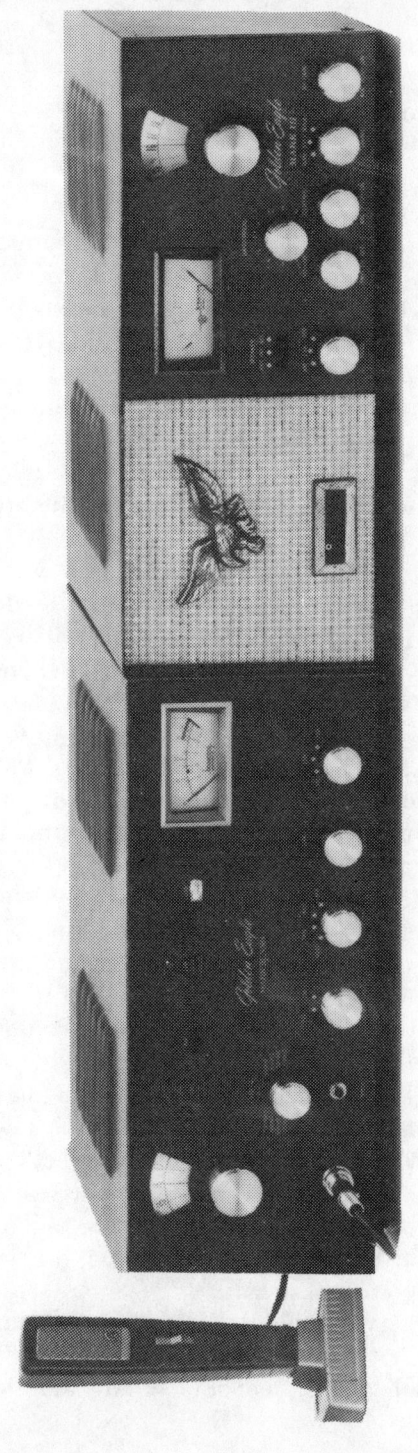

This equipment, produced by Browning Laboratories, is designed exclusively for base station use.

units are available, depending on whether you plan to power your unit with an auto battery, 117V AC house current, or nickel-cadmium cells.

BASE STATION / MOBILE TRANSCEIVERS

When a Citizens Band transceiver is to be installed in a home or office (called a "base station"), or in a vehicle or boat (called a "mobile"), most users select a unit with a 4-watt rating, the most powerful type allowed by the FCC. Only if cost is a factor should you consider purchasing CB transceivers using electron tubes over transistorized units. Transistorized equipment is more reliable since there are no fragile glass tubes to break or generate destructive heat. Transistorized "gear" is especially desirable for mobile installations because it has a much lower battery drain.

Incidentally, if you plan to purchase a unit for mobile use only, it makes little sense to pay extra dollars for a CB transceiver which permits operation from house current and auto batteries. Many manufacturers have accessory AC power units for home use if you need it. And if you plan to use only one or two channels, you won't want to pay $40 to $75 extra for a transceiver which comes equipped with all 23 channels. Only if you have need for monitoring all 23 channels, should you purchase a unit with tunable receiver, which costs as much as $35 to $60 additional.

Most users find that a unit with five or six switched channels are the easiest to operate. These normally have only three controls: an on-off-volume control which is exactly like the one on your TV set, a "squelch" control to cut out static and weak stations (which works much like the volume control), and a channel selector, just like the one on your TV receiver.

Nearly all 4-watt transceivers on the market today have "superhet" type receivers. However, the 4-watt transceiver you select should have a "double-superhet" or "double-conversion" receiver, which is more sensitive and selective than the "single-conversion" variety. The "single-superhet" or "single-conversion" type is good enough for walkie-talkies but not for more powerful sets.

Avoid transceivers which look as if the components were "poured" in, or those with flimsy cases. Either kind will not withstand much abuse, especially for mobiles.

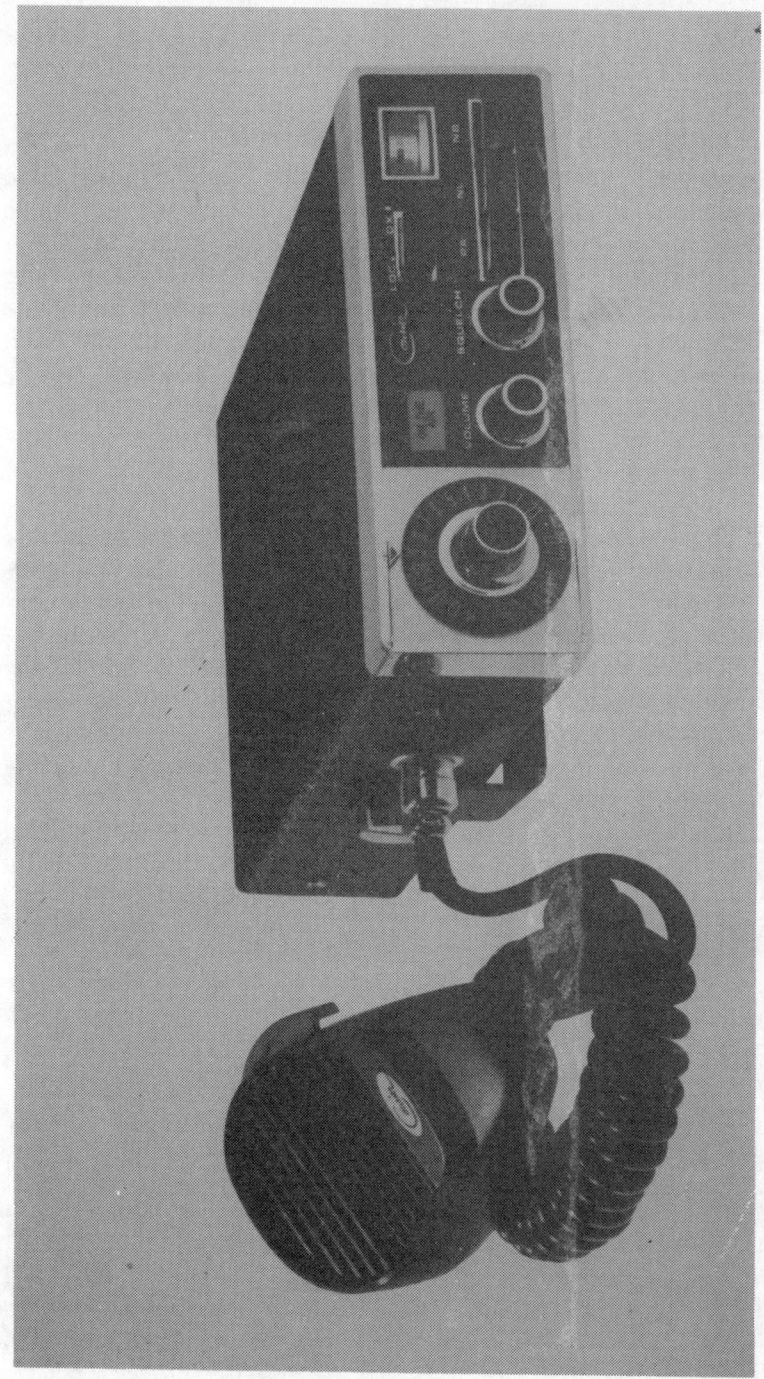

Courier 23-channel heavy-duty CB transceiver.

good base station transceiver is about $180 and about $10 less for a good mobile unit.

HAND-HELD WALKIE-TALKIE TRANSCEIVERS

Selecting a hand-held "walkie-talkie" type citizens band transceiver can be especially confusing because in addition to dozens of American-made units you will find hundreds of imported brands. To the casual observer there may appear to be few differences between units costing $15 apiece and those costing $80 or more. The cheapest kinds of Citizens Band hand-held units are little more than toys. They have transmitting ranges of only shouting distance—100 to 200 yards. And their receivers are so unselective that stations on virtually any channel will blot out communications on the channel to which the inexpensive unit is tuned.

If you are looking for a license-free one-tenth watt pair of transceivers, make sure that the units you select have "superheterodyne" or "superhet" type receivers. This is the only kind of receiver which is selective enough to discriminate between signals from units on other channels and signals on the channel to which your transceiver is tuned. Look at the specifications and check the receiver sensitivity rating—it should be one microvolt or better if you are to maintain communications over a one- to three-mile range.

If you wish to talk over a distance of three to five miles, hand-held units with 1.5- to 4.0-watt power ratings are avail-

High power hand-held units such as this Radio Shack TRC 101B include 23-channel operation.

50

able. They are the most desirable even over shorter distances since their signals will be much louder than those of one-tenth watt transceivers. They require an examination-free FCC license and can be used to communicate with more powerful units: for example, between a motorboat and a 4-watt base transceiver in a lakeside cottage. Again, these walkie-talkies have superheterodyne receivers with sensitivity ratings of at least one microvolt. In addition, you should look for a "selectivity" rating of at least 40 decibels (abbreviated dB), signifying that stations on channels adjacent to the one your unit is using will be adequately reduced in strength.

Other desirable features for such 1.5- to 4.0-watt hand-held transceivers are: rechargeable nickel-cadmium batteries (so you don't spend a fortune in flashlight cells), a "squelch" control for cutting out static and stations weaker than your own using the same channel, and provisions for connecting to a base station antenna.

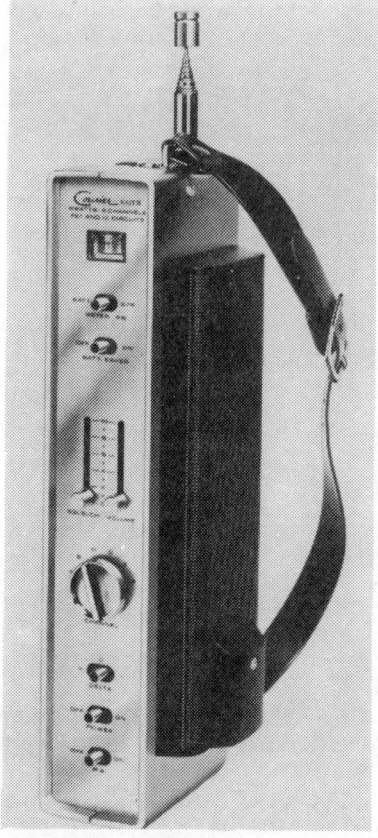

There almost seems to be no limit to miniaturization. This courier hand-held unit packs the punch of the full legal power limit and has six-channel capability, frequency vernier, squelch, S-meter, and other features.

Cobra 135 SSB/AM base station includes built-in digital clock with automatic equipment turn-on feature, SWR metering, and PA output. (Courtesy Dynascan Corp.)

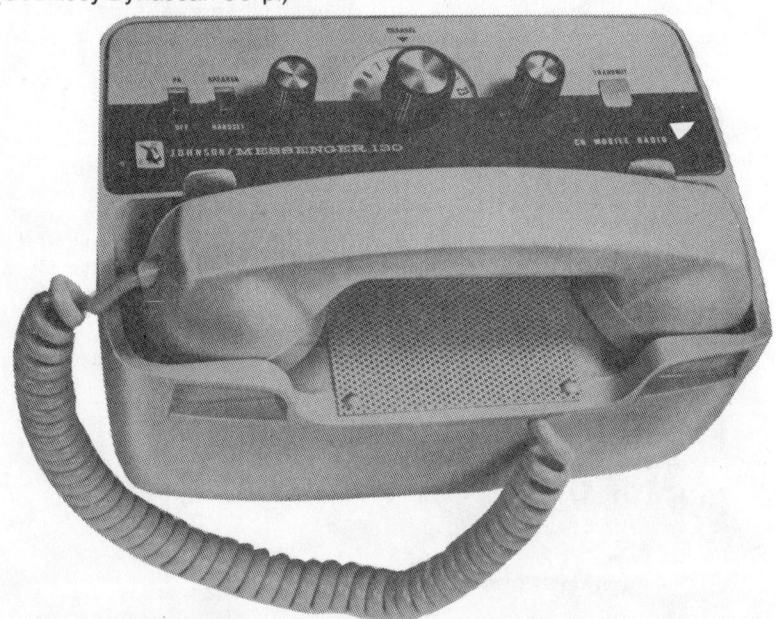

Messenger Model 130 mobile transceiver with telephone styling and a universal mounting arrangement provides a variety of installation configurations. (Courtesy E.F. Johnson Co.)

Hy-Range I, 23 channel mobile unit with a built-in mike preamp and automatic modulation control operates from either positive or negative ground. (Courtesy Hy-Gain Electronics Corp.)

Messenger 132 soid-state base station utilizes telephone handset type operation with automatic speaker silencing when the handset is lifted to give the option of private listening. (Courtesy E.F. Johnson Co.)

Hy-Range IV, 23 channel base station with continuous delta tuning, S/RF meter, variable antenna and load controls, ANL, built-in TVI filter and operational receiver VFO. (Courtesy Hy-Gain Electronics Corp.)

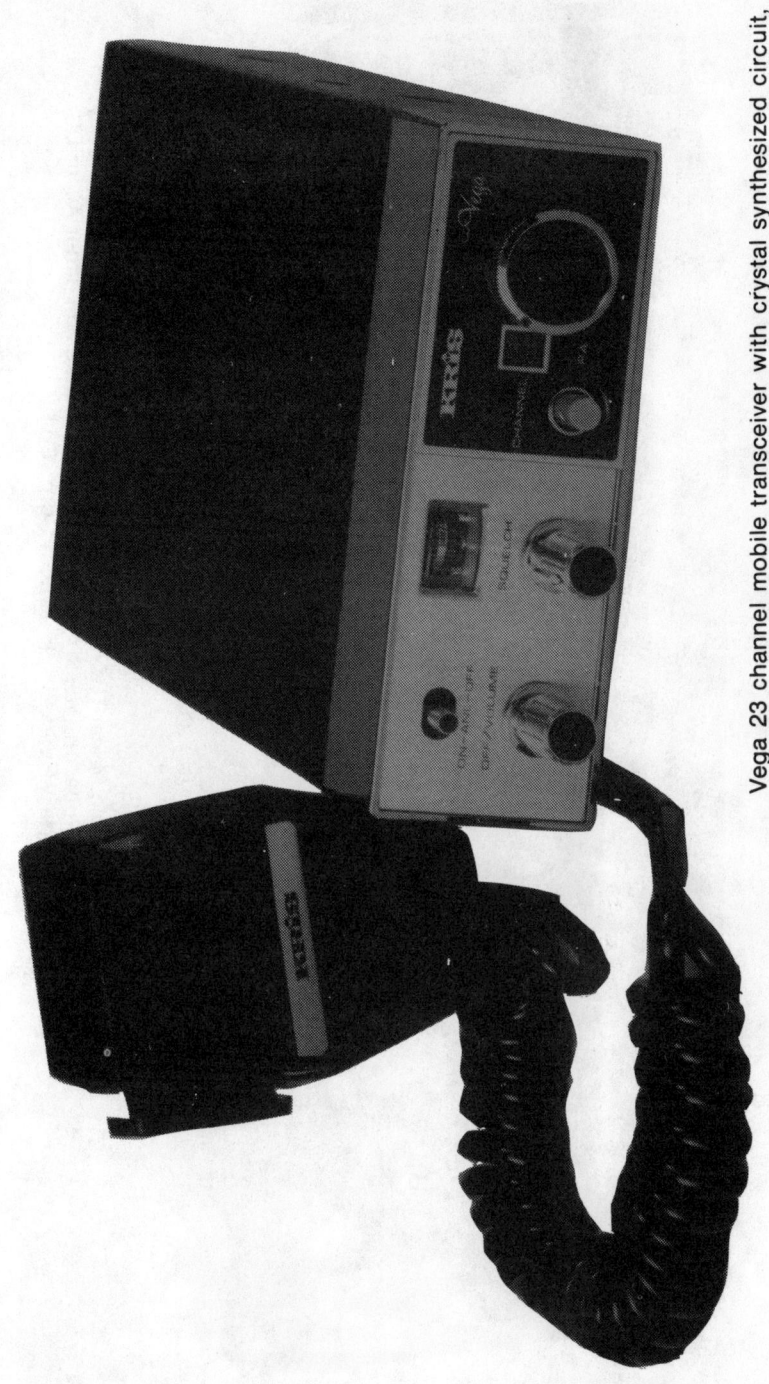

Vega 23 channel mobile transceiver with crystal synthesized circuit, S-meter and ANL can also be operated from 115 VAC source with optional AC power supply. (Courtesy Kris, Inc.)

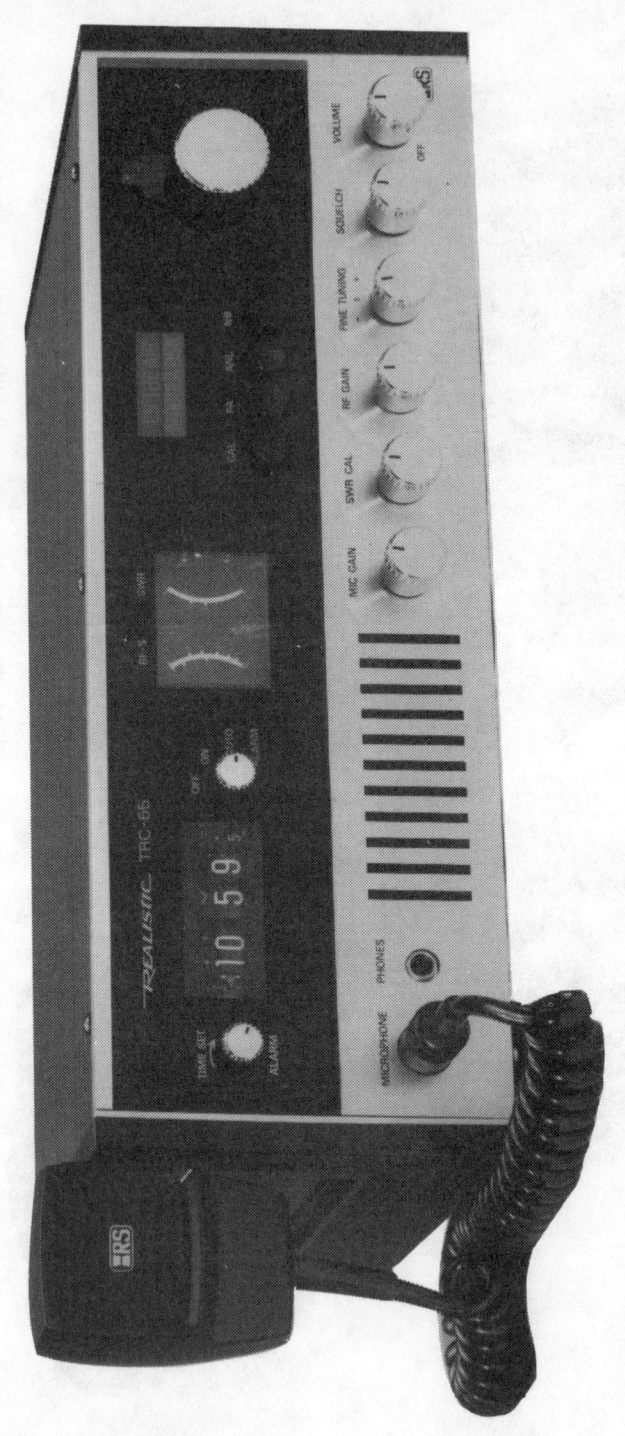

Realistic Model TRC-55 base station with digital clock and three-way delta tune, on-the-air modulation lights, operates on either AC or DC. (Courtesy Radio Shack, a Tandy Corp., Co.)

Model 606 mobile transceiver for positive and negative ground systems has unique remote switch control on the microphone to allow antenna channel switching. (Courtesy Royce Electronics Corp.)

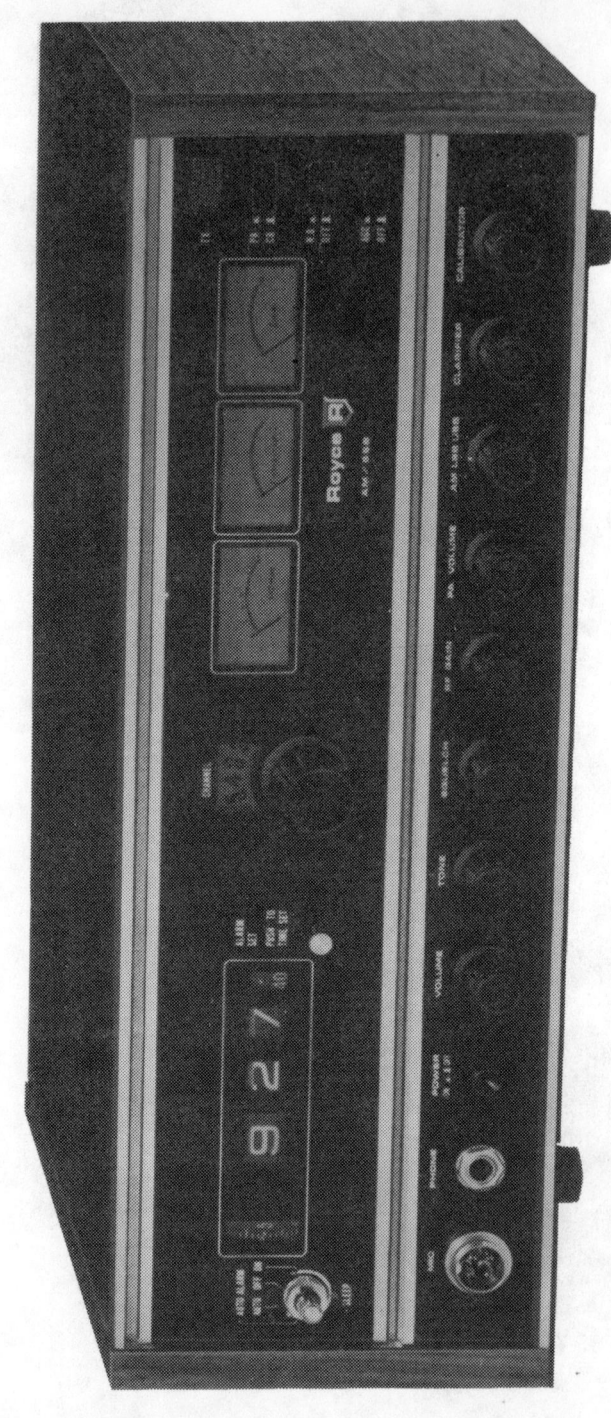

Model 1-620 23-channel AM base station for 115 VAC or 12 VDC positive or negative ground. (Courtesy Royce Electronic Corp.)

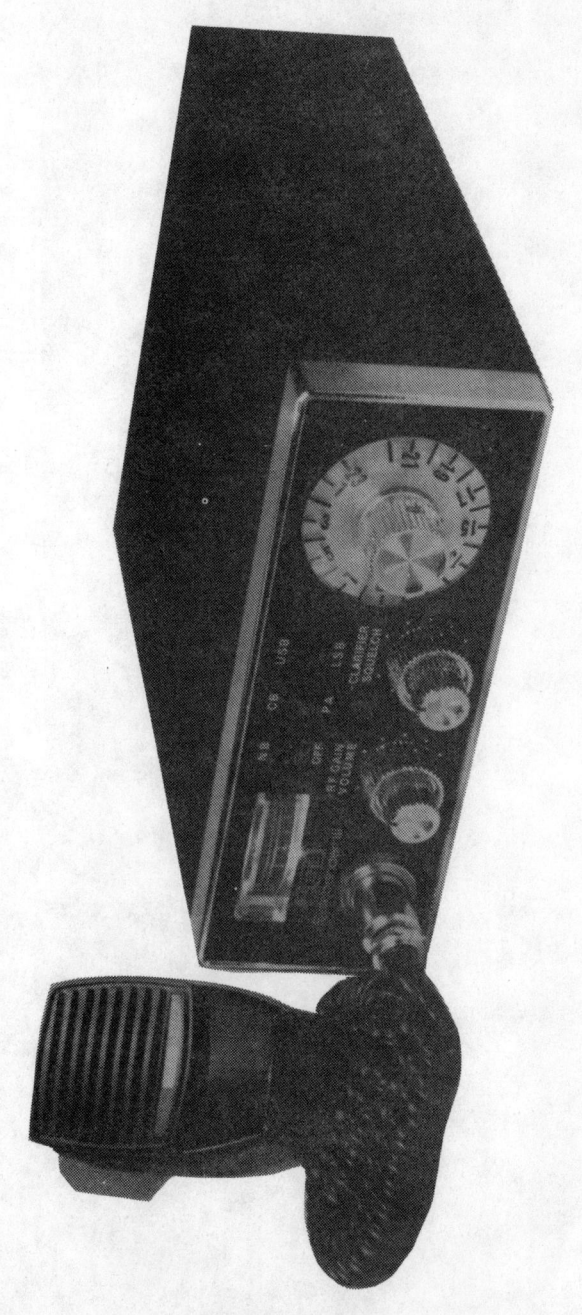

Sidebander III SSB mobile transceiver features 25 watts peak power on sideband with clarifier control for exact tuning and PA/Hailer provisions. (Courtesy SBE Linear Systems, Inc.)

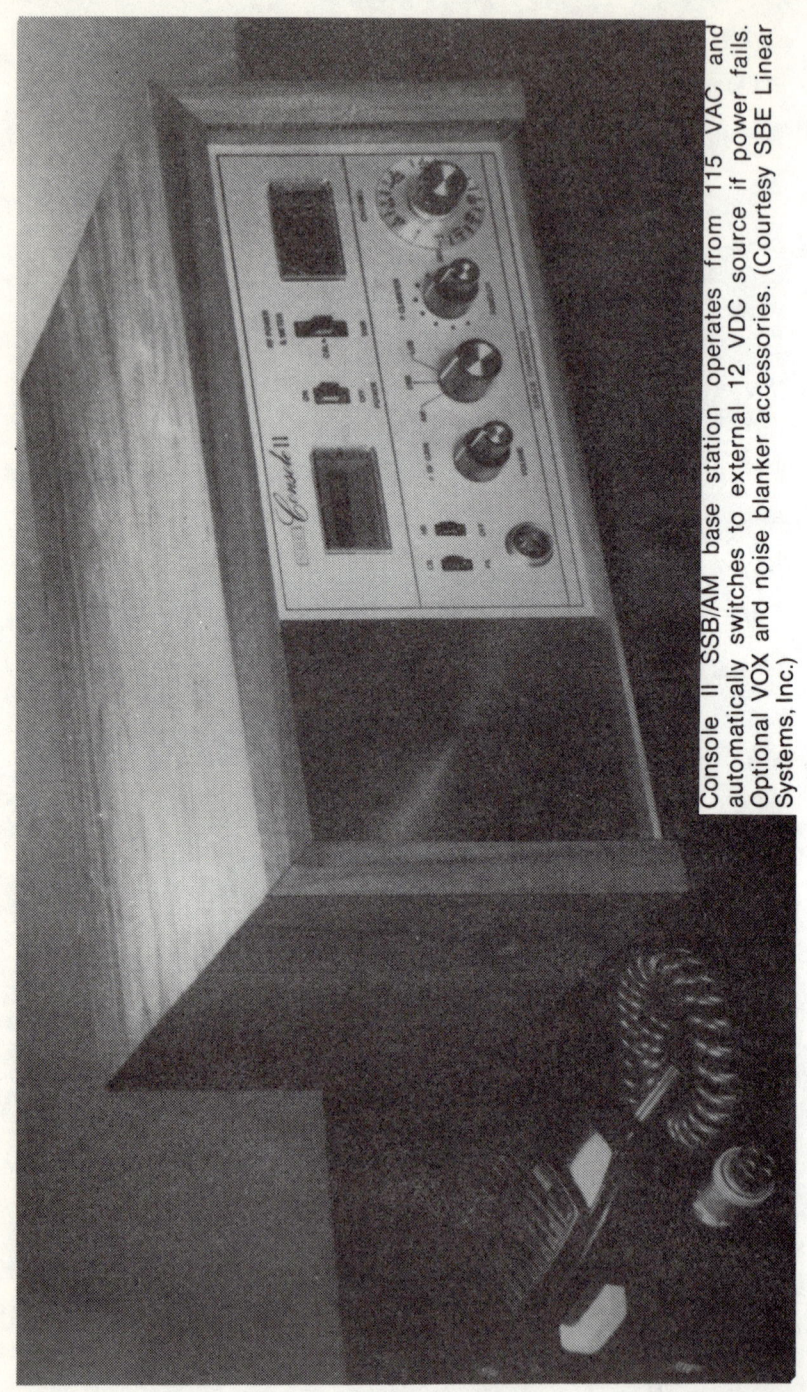

Console II SSB/AM base station operates from 115 VAC and automatically switches to external 12 VDC source if power fails. Optional VOX and noise blanker accessories. (Courtesy SBE Linear Systems, Inc.)

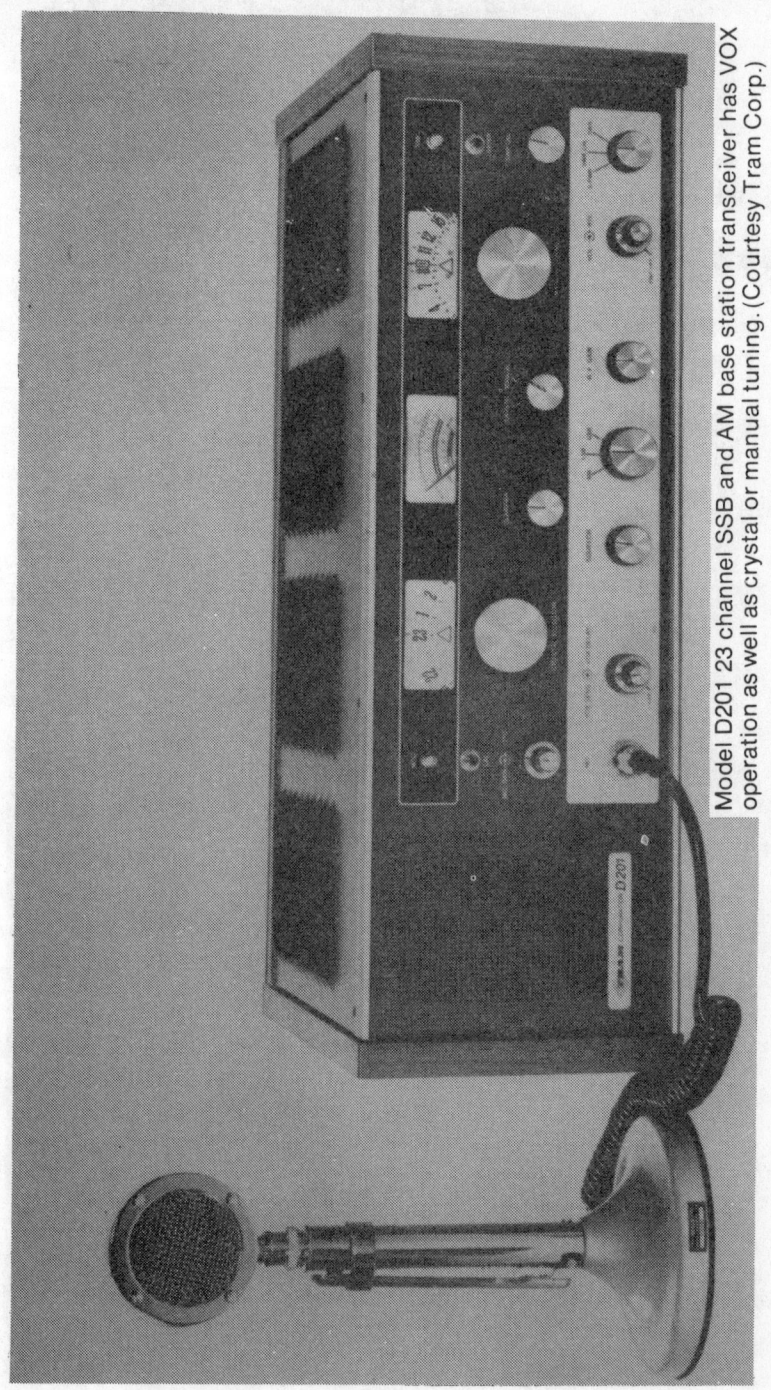

Model D201 23 channel SSB and AM base station transceiver has VOX operation as well as crystal or manual tuning. (Courtesy Tram Corp.)

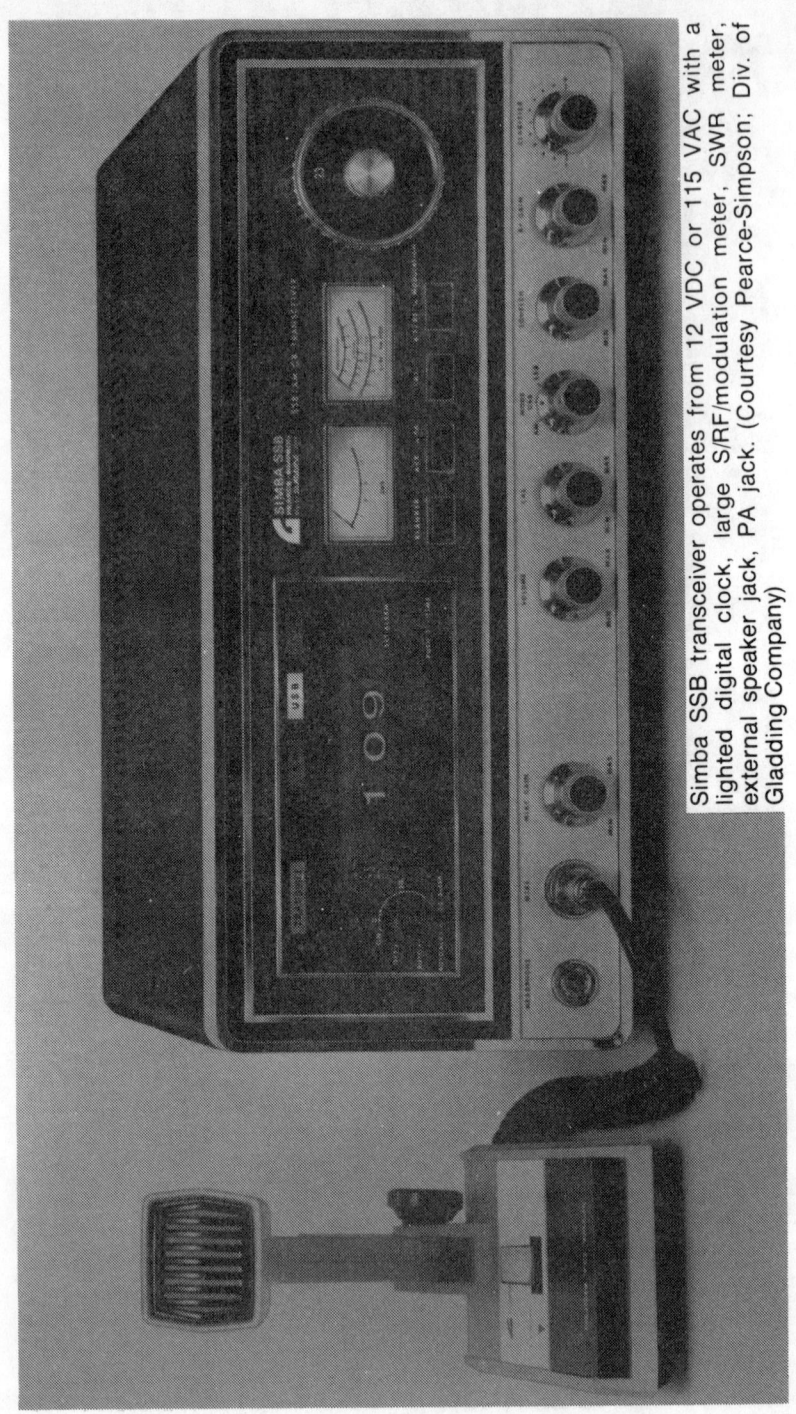

Simba SSB transceiver operates from 12 VDC or 115 VAC with a lighted digital clock, large S/RF/modulation meter, SWR meter, external speaker jack, PA jack. (Courtesy Pearce-Simpson; Div. of Gladding Company)

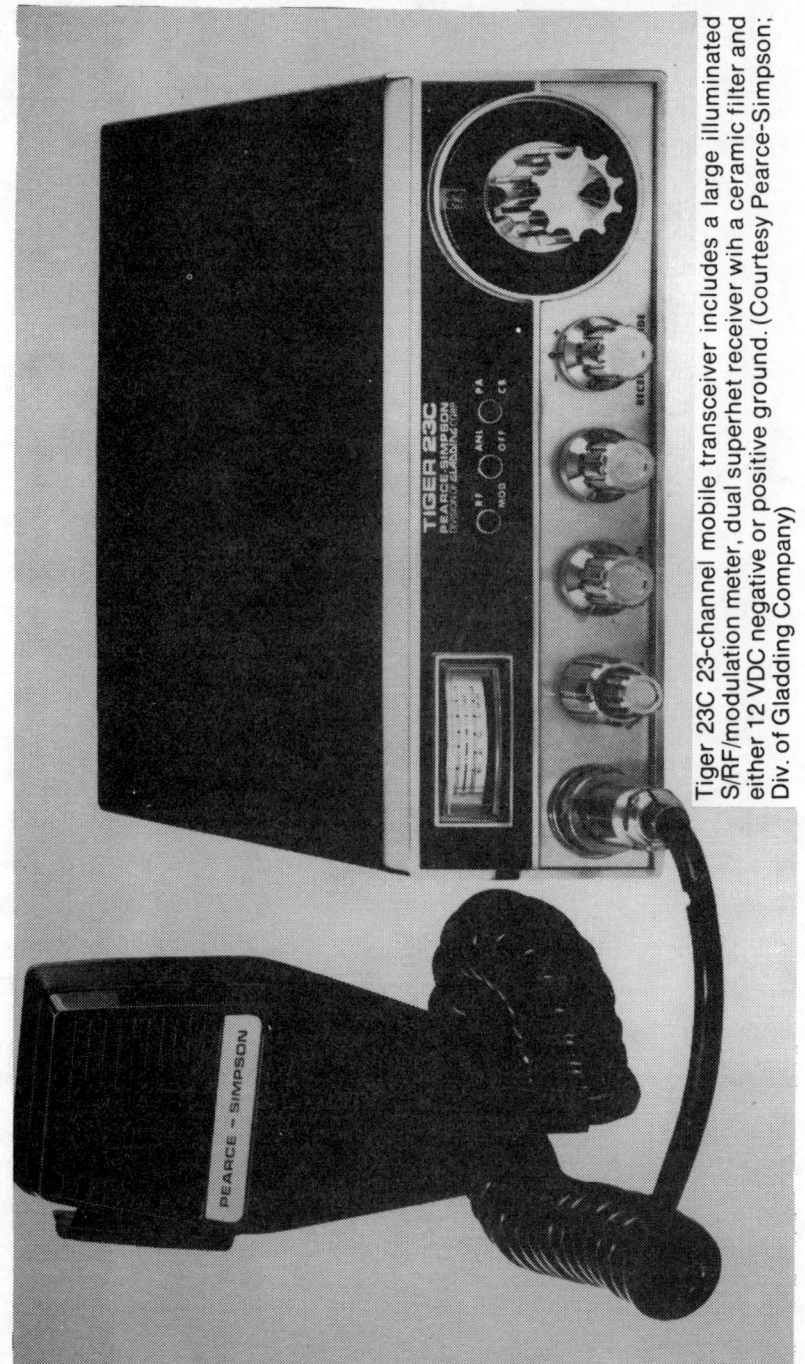

Tiger 23C 23-channel mobile transceiver includes a large illuminated S/RF/modulation meter, dual superhet receiver with a ceramic filter and either 12 VDC negative or positive ground. (Courtesy Pearce-Simpson; Div. of Gladding Company)

LTD SSB/AM mobile transceiver combines gold and walnut grained exterior in a compact package with a tamper-proof mounting system. (Courtesy Browning Labs, Inc.)

Com-Phone 23, 23 channel transceiver designed for mobile or base station operation (with operational AC power supply) features automatic noise limiting, AGC and PA. (Courtesy Lafayette Radio Electronics Corp.)

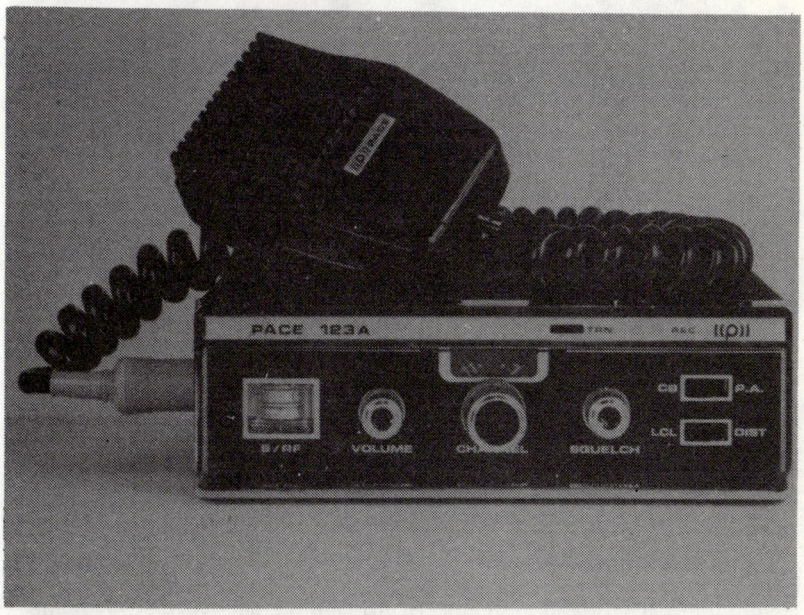

Model 123A mobile unit for operation from either positive or negative ground includes a transmitter indicator light, S-meter/modulation indicator and PA capability. (Courtesy Pace Div., Pathcom, Inc.)

Here's one two-way CB radio that puts all the knobs you need in the palm of your hand. It's the Radio Shack TRC-61 One-Hander, has on-off/volume, squelch and 23-channel selector built into the microphone. The remote section can be mounted out of sight in a concealed location as a deterrent to theft.

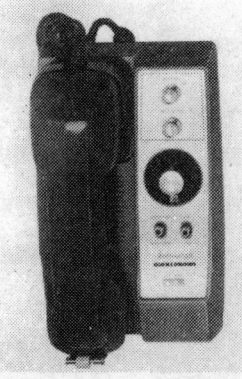

Dolphone marine CB transceiver with functional handset to allow communication under noisy conditions features 23 channel operation, ANL and PA. (Courtesy Unimetrics, Inc.)

BUYING USED CB RADIO GEAR

Buying used CB gear can be fun if you are inclined to be a "swap artist" or "horse-trader." About all you need to qualify in this category is a piece of used gear you want to get rid of and someone or some place to trade with. A local dealer may be a good choice, especially if he's in the habit of taking used equipment in on trade.

There are some basic ground rules for this art, just as in trading cars. Be sure you clean your rig up. Clean up the cabinet. Touch up any scratches or chip marks if necessary and clean out the dust. If the rig doesn't work and needs only minor repairs, either have it fixed or explain it to the prospect. It helps to provide any manuals or schematics with the unit, too.

When you trade for a better piece of gear, you probably will be a little disheartened at the allowance you get. It's just one of the facts of trading. Remember, a dealer expects to make a profit on the sale of your equipment and if it's an obsolete piece of gear, he has to consider the time it will be on the shelf. You may do better trading for a new transceiver, because the margin for profit is better.

It's easy to get subjectively involved with your old gear; chances are it will seem far more valuable to you than to the next fellow. And remember, prices for CB gear are continually dropping—the unit that cost several hundred dollars a few years ago might sell for half that today, even though it might never have been used at all.

Some of the items to look for in the transceiver you wish to purchase or trade for are as follows. These may vary somewhat, but will serve as basic guidelines:

Hand-held transceivers, such as one being discussed above in author's office, are at their best in the field—where neither automobile voltage or household 117V AC are available.

Used equipment is frequently offered at attractive prices in CB outlets. The above display is at Allen Electronics, Flushing, New York.

Cobra 28 mobile transceiver with a "scan alert" system which allows for monitoring and transmitting on specific channel other than the one selected by the channel selector. (Courtesy Dynascan Corp.).

A

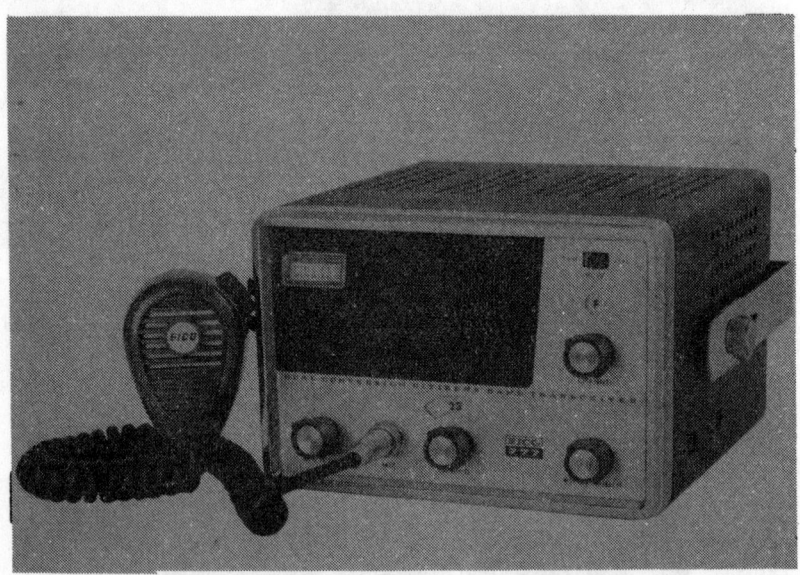

B

Always be on the lookout for "grand opening" sales, where substantial savings can be had on new equipment if you get there in time! Reconditioned CB equipment, such as the vintage EICO Model 777 transceiver shown here, can be real bargains to knowledgeable buyers who know just what to look for.

1. Will it provide the additional functions you want?
2. Does the price seem justified?
3. Is it commercially built or from a kit?
4. Does it need repair?
5. Is a schematic available?
6. Is there a guarantee?
7. Are replacement parts available?

In most cases the dealer may want to run a test of your equipment. You have every right to suggest the same courtesy.

SSB EQUIPMENT

Conventional AM (amplitude modulated) transmission, the type we normally find in CB equipment, divides its power into the "carrier" and "sidebands." There are two sidebands, upper and lower. The "carrier" is the transmitter channel operating frequency. This "carrier" is modulated by audio when you push your mike button and talk. Your voice is amplified and combined with the transmitter output signal to vary or "modulate" it. The actual voice information is in the sidebands, and the carrier itself is nothing more than excess baggage. So why not eliminate the carrier?

In SSB, or single-sideband transmission, the carrier is eliminated. In fact, so is one of the sidebands. Since the power in AM transmitters is divided into the carrier and two sidebands, it makes sense that by eliminating the carrier and one sideband, more power can be concentrated in the remaining one. That's one of the advantages of SSB operation, the transmitter can make better use of its power. Another advantage of SSB operation is that it is not as affected by interference as conventional AM rigs are, and SSB signals are more readable on crowded channels. The disadvantage of SSB operation presently is cost. If you purchase SSB units to complete the system you want, you will be able to communicate only between the SSB transceivers. On the other hand, none of the AM units will be able to monitor your signals—at least not with any great degree of intelligibility.

Chapter 4

Hand-Held Transceivers

If you've heard a lot about walkie-talkies recently, it's for good reason. One of the controversies in the CB field concerned itself with whether or not the gadgets should operate on their own "exclusive 49 MHz band" or continue functioning on CB frequencies. Meantime, walkie-talkies are selling more than even before and more and more manufacturers are hopping on the bandwagon.

Electronic financial moguls figure that the pushbutton wonders outsell conventional CB gear more than five to one in terms of total annual volume. Predictions for 1976 are that nearly one billion dollars worth of the devices will be sold — and this does not include the other types of walkie-talkies being used extensively by municipalities, utilities, police departments, etc. No one seems to know if the feared "saturation level" will ever hit this market. Right now it is clearly the biggest money-maker in the entire two-way radio communications industry.

It's not hard to understand the popularity of these gadgets: They're compact and portable, they really do work, and they're relatively cheap. Pass any corner discount store these days and you'll be greeted with a window display just brimming over with glistening, chrome-antenna'd walkie-talkies — many priced as low as $9.88 per pair. On the other hand, the major CB manufacturers are churning out $99-per-pair high-powered units which make these peanut whistles look like toys and boast of "ranges up to 20 miles." Before you rush out and plunk down your hard-earned dollars for a pair of these electronic marvels, however, let's consider just what you can do with them and what you'll really be getting for your money.

Basically, there are two types of walkie-talkies that operate on CB frequencies: (1) the Class D (must be licensed by FCC) models, and (2) the so-called "license-free" low-power units, which have come to be known as "Part 15" units.

This simplistically styled walkie-talkie has the power of a base station as well as some of its features.

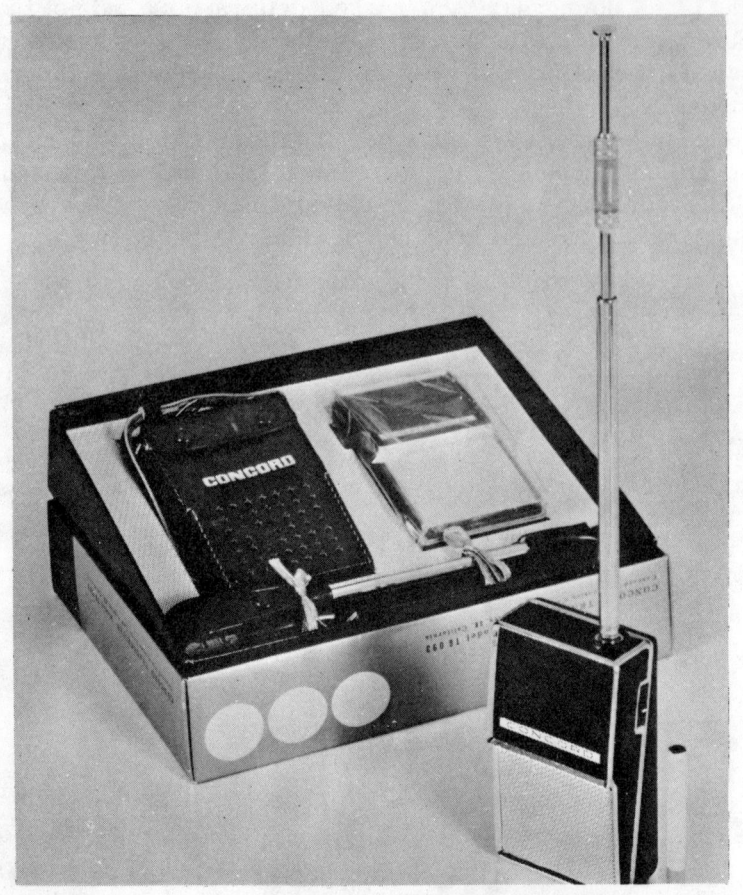

More elaborate is this set, available through department stores, discount houses, and electronics outlets, produced by Concord Electronics. Notice the center-loaded whip antenna.

WHAT FCC SAYS ABOUT "NO LICENSE" TRANSCEIVERS

Most walkie-talkies sold over-the-counter these days fall into the last category, the no-license types. The nickname referred to earlier has come from Part 15, a regulatory section of the FCC Rules which pertains to "non-interference" communications. To qualify for license-free operation under this open-end clause, a transceiver can be designed for a wide range of frequencies (the AM wireless broadcasters, for example) so long as its power input does not exceed 100 milliwatts (1/10th of a watt) and it is legally operated. Even more attractive is the fact that there are no age requirements for the user, who might otherwise not be permitted to communicate on CB frequencies, since you must be 18 years of age or older for a Class D ticket.

Because anything under one-tenth watt qualifies for Part 15 does not imply, however, that all units are rated at 100 milliwatts or are even investigated by a governmental agency. The sad truth of the matter is that the FCC imposes no design or construction restrictions upon the manufacturers—only on the user, caveat emptor. The FCC does ask Part 15 walkie-talkie enthusiasts, though, to observe a few basic ground rules; for the most part, however, they are not as sticky as those for licensed communications. Since they're looked upon as "non-interference" units, the user must relinquish the

Anxious to call in youngsters for lunch, this farmer's wife is careful to listen first until the existing conversation on the frequency is terminated. Always remember: walkie-talkie communications are not private; listen first to avoid accidentally jamming another's message.

The Part 15 unit this youngster is using (yes, it's all perfectly legal) cost approximately $50 per pair. Performance characteristics, of course, greatly exceed those of toy-store-variety units. Notice the "good/bad" battery meter on hand-held transceiver at right.

channel if his operation is interfering with communications of a Class D user.

The general rule is to listen first (to insure that no one is on the air) and then, once you've begun your conversation, periodically check the channel. If you consistently cause problems for a licensed CBer, he can report you to the FCC. Further, the law demands that you don't transmit profane or indecent language, that you don't broadcast phoney emergency or distress messages, and that you don't slander anyone over the air. Additionally, intermittent pressing-down of the transmit button is considered to be malicious interference and con-

strued as "jamming," punishable under federal law by a stiff fine.

The most convenient way to avoid problems with Part 15 regulations is simply to remember that your walkie-talkie communications aren't private. They can be heard by anyone who also has a walkie-talkie on the same channel, to say nothing of eavesdropping licensed CBers. It's like a long-distance PA system, with a diverse audience of listeners. Bear this in mind, and chances are you'll never have to worry about breaking the FCC rules. Operationally, Part 15 people can communicate only with other unlicensed walkie-talkies. Under no circumstances may you talk with, for example, a 5-watt Class D licensee. Not only would the CBer be risking his license, but you would stand a stiff fine to boot.

EMERGENCY OPERATIONS ON CITIZENS RADIO CHANNEL 9 (27.065 MHz)

Each licensee must determine if his message is a valid emergency communication. An emergency is defined as involving the **immediate** safety of life or the **immediate** protection of property. Examples are listed below:

EMERGENCY	EXAMPLE MESSAGE
Yes	"A tornado sighted six miles north of town."
No	"This is observation post number 10. No tornados sighted."
Yes	"I am out of gas on Interstate 95."
No	"I am out of gas in my driveway."
Yes	"There is a four car collision at Exit 10 on the Beltway, send police and ambulance."
No	"Traffic is moving smoothly on the Beltway."
Yes	"Base to Unit 1, the Weather Bureau has just issued a thunder-storm warning. Bring the sailboat into port."
No	"Attention all motorists. The Weather Bureau advises that the snow tomorrow will accumulate 4 to 6 inches."
Yes	"There is a fire in the building on the corner of 6th and Main Streets."
No	"This is Halloween patrol unit number 3. Everything is quiet here."

ADVANTAGES OF PART 15 TRANSCEIVER OPERATION

With virtually no restrictions on hobby use of non-Class D walkie-talkies, many fellows have formed flea-power DX clubs. These groups specialize in seeing how far they can communicate with Part 15 walkie-talkies. Indeed, there have been instances of contacts over several hundred miles. A few

This Chicago lad attempts conversation with another Part 15 operator 700 miles distant. While unusual, these contacts are possible through a propagational phenomenon known as sporadic-E skip.

years ago, in fact, there was a case of a 1400-mile contact due to a propagational phenomenon known as sporadic-E skip. Since these "freaks" have been occurring more and more frequently in recent years, the DXers are even converting to CW (code) operation, based on the fact that code will cut through the distance barriers much more effectively than regular voice modulation. Caught in a pinch, many enthusiasts tap out International Morse Code by pressing their trans-

Best time for long-range local (ground wave) communications is late at night, as this explorer scout on field expedition knows well. Additionally, every added foot in height increases communications range appreciably.

mit-receive button! Many hobbyists delight in "converting" their Part 15 rigs to base stations, complete with code keys, headsets, and even AC power supplies to replace their battery packs. General Electric brought out a commercial version of such a base setup which looks a great deal like a ham radio station, yet it's all perfectly legal.

Since you'll be working with a telescoping mast whip antenna and much less power than your Class D counterpart, your range expectations will be somewhat lower. Best bet is to ignore the "one-mile communications" propaganda manufacturers churn out and figure your dependable two-way range as considerably less. Under normal conditions, and operating with a unit that has a superheterodyne receiver, you can count on several city blocks, in open country (pretty much line-of-sight, no Empire State Buildings in the road, etc.) a lot further. Across open water is generally the best bet: frequently you'll get 5 miles or more.

Forget about hard-and-fast range estimates, and count on a minimum communications range; when you exceed this, it'll be all the more exciting. Best time to get greatest local distance is in the wee hours of early morning, when there's hardly anybody on, and when the proverbial "night-time factor" is in full swing. It's a known fact that after midnight you can often communicate over three, four, and five times the normal non-skip distances. If you're free of interference, you can also take advantage of a multitude of daytime phenomena —such as tropospheric bending (which will get up to 25-35 miles), tropo ducting (up to 75-125 miles), and the elusive Sporadic-E which likes to pick up your signals and deposit them 500-700 miles distant, and frequently <u>double</u> that!

Additionally there is the fact that you can say just about anything you want over the air providing you don't go <u>too</u> far out. For example, you can broadcast ham-type "CQ's" if you like. Additionally, there is no time-limitation on how long you can talk (such as the 5-minutes-on, 1-minute-off Class D recommendations). And if you want to tamper a bit with your transceivers, you don't need an affidavit from the FCC stating that you're a 1st or 2nd Class Commercial licensee in order to do it. If you're a really eager no-license enthusiast, here's a tip: try to get a pair of sets equipped for Channel A (27.995 MHz), the hobbists' favorite frequency. If you're a CW eager-

Crew foreman contacts a worker on the 24th floor of a NYC building with Part 15 unlicensed transceiver. Tall buildings such as shown here, however, keep communications range to a minimum—unless you're on one!

beaver, switch to Channel B (27.045 MHz). And if you really want to go all-out, you can wire in a variable-frequency oscillator (VFO) which will allow you to use any of the frequencies on the 26.970-27.270 MHz band without crystals!

AVAILABLE TYPES OF PART 15 UNITS

There seems to be an almost endless stream of different brands of walkie-talkies on the market these days. They aren't all alike, something you probably discovered while you were looking at the $7.98 units and compared them with look-alikes going for $29.95. For the most part, the bulk of the cheaper units are intended to be used as simply toys. These types are characterized by superregenerative receivers, plastic cases, single-channel construction, three or four transistors, and actual power levels as low as 30 to 50 milliwatts (under 1/20th of a watt). While these sets are most certainly worth the $10 or so cost, they should hardly be considered as a reliable means of communications by the serious user. For the kiddies, you can't beat them; a youngster will get just as big a charge from a $6.00 unit as he would from one costing $25. Additionally, the end result will be much the same after he's used the telescoping mast as a baseball bat or has a few antenna duels, when the family car's been driven over it in the driveway, or when he's taken it in the bathtub to test its underwater effectiveness. Even if you have the most extraordinary youngster in mind, he'll quickly lose interest when he realizes that it's his allowance money that has to be paid for new batteries when the set's gone dead and the novelty's worn off. Sets that sell for more than $15-$16, however, are what you are after if you figure you'll ever really want to use them for actual radio communications. You'll have to pay this much for a superheterodyne receiver, essential if you want to hear the other end of the conversation. Further, these sets have a bit more transmit pizazz than their cheaper counterparts.

As prices increase, you'll notice that various added feature attractions are incorporated into the models along the way. Most of the better types have a "tone-signaling" feature—a button that you can push to signal a distant station. A high-pitched tone tells your companion station that he'd better get to his unit for an in coming message. Many of these same types feature "lock-in" transmitting, so you won't wear your thumb to the bone fighting the push-to-talk button's mainspring during long transmissions.

A few more goodies to keep an eye out for in Part 15 sets include: range boosting in the audio circuits, multi-channel operation, a squelch control, 10 or more transistors overall, good noise limiting for high-noise-level areas, push-pull receiver audio, separate mike and speaker, and even provision for an external 117v AC power supply and/or battery charger. The "ultimate" Part 15 walkie-talkies even include

On this southern Illinois farm, both four-watt transceivers and walkie-talkies are used to direct movement of men, equipment and livestock. Handheld transceivers are one-watt walkie-talkies licensed as Class D units.

a battery-condition meter that tells you ahead of time whether or not to charge or replace the cells!

Remember, however, that the maximum power in any unlicensed set is going to be 100 milliwatts. And no more. Once this simple objective has been achieved, start looking to the receiver for added extras. Here you'll be getting exactly what you pay for. Many of the extremely cheap sets look darned pretty from the outside, but don't buy until you find out what's happening inside. By the same token, certain unscrupulous overseas producers are packing many Part 15 units with "16 transistors." Close inspection may reveal that only seven are working in the circuit. The other nine are either mere duds or are connected in superfluous configurations.

HAND-HELD CLASS D CB TRANSCEIVERS

Sooner or later for true high-power operation you'll want to "graduate" from 100-milliwatters to something a bit higher powered. Here again, there are a host of units to choose from, priced anywhere from $29.95 to more than $100. Structurally, these sets are a bit more hefty than their Part 15 counterparts, and for good reason: Most are constructed with several times the "innards," and they're generally built to take a beating. Largely produced by American companies for the more sophisticated user, most buyers are already licensed CBers with elaborate base stations. For these chaps, the walkie-talkies are used for direct communication with base stations. On the other hand, these high-power units are also many people's first real experience with 27-MHz CBing, and it can be a quite satisfactory one indeed.

While you must pay an $4.00 filing fee to the Federal Communications Commission and fill out a temporary (form 555B) license before you go on the air , you'll invariably find your walkie-talkie already packed with the FCC Form and full instructions. The manufacturer usually instructs you on how to fill out the form and supplies plenty of reading material to acquaint you with CB while you're waiting for that license.

There are nearly 100 different models of Class D sets to choose from that take the form of walkie-talkies, and nearly all are worth every penny of their cost. Unless abused, they'll outlive you. Bear in mind that many of these sets perform better and have more features than you could buy in even a base station a few years ago—at any price! They owe their existence to the fantastic technological strides made recently: the advent of mass-produced field-effect transistors and extremely complex integrated circuits.

Most of the high-power sets are solidly built and backed up by warranties. Once again, however, you're going to get exactly what you pay for. Consider first the almighty dollar and how much power you really need in your units, then go on to the "extras" such as battery-condition meters, range-boosting, etc. Many buyers seem to over-emphasize the appearance of their chromed sets, to the neglect of the actual power levels and communications capability. Frequently, the most ungainly-looking walkie-talkies are the best performers in all types of weather. A word of warning: Do not try to link these sets to your 100-milliwatt Part 15 setup. Just because they look similar doesn't imply you can talk to anyone other than licensed Class D base, mobile, or portable stations. There are stiff penalties for ignoring this!

THE IMPORTS

Most hand-held sets—particularly the Part 15 unlicensed types—come in from Japan and the importers put their own brand names on the units. It's not unusual for the very same walkie-talkie to be selling under four or five different brand names...and at different prices! This isn't to knock the performance or quality of the sets, however; it's just to let you know that the brand name doesn't necessarily tell you where the radio was manufactured. But don't worry about it; a Japan-built walkie-talkie is quite apt to be superior to a similarly priced U.S. model.

Always get full information on the back-up guarantee of the sets you intend to buy. You don't want to be mailing your $12.95 walkie-talkie back to Japan for repairs. If you can't have the guarantee apply for servicing here in the United States, forget the whole thing.

Watch for walkie-talkies selling for "50% off" in those going-out-of-business novelty shops. You'll never get them serviced, and invariably there's something wrong with the sets in the first place. Additionally, intelligent comparison shopping frequently will reveal that a comparable guaranteed set can be had by mail from Lafayette, Olson, or Radio Shack for the same price—and less! Your best bet is always to rely on reputable electronics suppliers and mail-order houses.

ENJOY YOURSELF

While selecting your unit is serious business, you can still have a lot of fun with your walkie-talkies—providing you adhere to the FCC regulations as they apply to your type of set. For Class D operations, you'll want to check the same information presented elsewhere in this book that pertains to the 4-watt licensed base stations. The exact same rules apply to all walkie-talkies running more than 100 milliwatts input. For the unlicensed Part 15ers, however, restrictions are quite light. In fact, you can make up your own "official" callsigns for use at outings, field trips, hunting or fishing expeditions, etc. Just make certain your callsign isn't similar to an established FCC ham or CB call. Something on the order of "N-880 calling N-881...Go ahead!" is fine and perfectly legal.

EXTRACTS FROM PART 15—RF DEVICES

§ 15.3 General condition of operation.

Persons operating restricted or incidental radiation devices shall not be deemed to have any vested or recognizable right to the continued use of any given frequency, by virtue of prior registration or certification of equipment. Operation of these devices is subject to the conditions that no harmful interference is caused and that interference must be accepted that may be caused by other incidental or restricted radiation devices, industrial, scientific or medical equipment, or from any authorized radio service.

§ 15.4 General definitions.

(a) *Radio frequency energy.* Electromagnetic energy at any frequency in the radio spectrum between 10 kHz and 3,000,000 MHz.

(b) *Harmful interference.* Any emission, radiation or induction which endangers the functioning of a radionavigation service or of other safety services or seriously degrades, obstructs or repeatedly interrupts a radiocommunication service operating in accordance with this chapter.

(c) *Incidental radiation device.* A device that radiates radio frequency energy during the course of its operation although the device is not intentionally designed to generate radio frequency energy.

(d) *Restricted radiation device.* A device in which the generation of radio frequency energy is intentionally incorporated into the design and in which the radio frequency energy is conducted along wires or is radiated, exclusive of transmitters which require licensing under other parts of this chapter and exclusive of devices in which the radio frequency energy is used to produce physical, chemical or biological effects in materials and which are regulated under the provisions of Part 18 of this chapter.

(e) *Community antenna television system.* A restricted radiation device designed and used for the purpose of distributing television signals by means of conducted or guided radio frequency currents to a multiplicity of receivers outside the confines of a single building.

NOTE: The television signals that are distributed are modulated radio frequency signals and may be:

(a) Broadcast signals that have been received and amplified.

(b) Broadcast signals that have been received and converted to another frequency.

(c) Any other modulated radio frequency signals fed into the system.

(f) *Low power communication device.* A low power communication device is a restricted radiation device, exclusive of those employing conducted or guided radio frequency techniques, used for the transmission of signs, signals (including control signals), writing, images and sounds or intelligence of any nature by radiation of electromagnetic energy.

EXAMPLES: Wireless microphone, phonograph oscillator, radio controlled garage door opener and radio controlled models.

§ 15.11 Prohibition against eavesdropping.

(a) No person shall use, either directly or indirectly, a device operated pursuant to the provisions of this part for the purpose of overhearing or recording the private conversations of others unless such use is authorized by all of the parties engaging in the conversation.

(b) Paragraph (a) of this section shall not apply to operations of any law enforcement officers conducted under lawful authority.

§ 15.63 Radiation interference limits.

(a) The radiation from all radio receivers that operate (tune) in the range 30 to 890 MHz, including frequency modulation broadcast receivers and television broadcast receivers, manufactured after the effective date specified in § 15.72 shall not exceed the following field strength limits at a distance of 100 feet or more from the receiver:

Frequency of radiation (MHz)	Field strength (uV/m)
0.45 up to and including 25	See paragraph (b).
Over 25 up to and including 70	32.
Over 70 up to and including 130	50.
130–174	50–150 (linear interpolation).
174–260	150.
260–470	150–500 (linear interpolation).
470–1000	500 (see paragraph (c) below).

(b) Pending the development of suitable measurement techniques for measuring the actual radiation in the band 0.45 to 25 MHz, the interference capabilities of a receiver in this band will be determined by the measurement of radio frequency voltage between each power line and ground at the power terminals of the receiver. This requirement applies only to radio receivers intended to be connected to power lines of public utility systems. For television broadcast receivers the voltage so measured shall not exceed 100 uV at any frequency between 450 kHz and 25 MHz inclusive. For all other receivers the voltage shall not exceed 100 uV at any frequency between 450 kHz and 9 MHz inclusive, 1000 uV for frequencies between 10 MHz and 25 MHz and linear increase from 100 uV to 1000 uV for frequencies between 9 MHz and 10 MHz.

(c) For television broadcast receivers the limit in the band 470–1000 MHz shall be 350 $\mu V/m$, compliance being determined as follows:

(1) Measurements shall be made at the following 10 frequencies in the band 470–1000 MHz.

MHz	MHz	MHz
520	700	850
550	750	900
600	800	931
650		

NOTE: If measurements cannot be made on one or more of the frequencies listed because of the presence of signals from licensed radio stations, measurements should be made on a nearby frequency. The report should indicate the actual frequency(ies) on which measurements were made.

(2) The average of the 10 measurements shall not exceed 350 $\mu V/m$.

(3) No measurement shall exceed 750 $\mu V/m$.

(d) Notwithstanding the provisions of paragraph (a) of this section and subject to the prohibition against emissions on the frequencies listed in § 15.215(c), the level of emission of RF energy from the receiver used with a radio control for a door opener shall not exceed the values listed below when measured in accordance with the procedures laid down in FCC Technical Division Report, T-7001, dated October 1, 1970.

Frequency (MHz)	Field strength at 100 ft. ($\mu V/m$)
Over 25 up to and including 70	32.
Over 70 up to and including 200	50.
200–1,500	50–500 (liner variation).
Over 1,500	500.

[§ 15.63(d) added eff. 5–7–71, this date later stayed to 11–1–71; II(69)–71]

SUBPART E—LOW POWER COMMUNICATION DEVICES

§ 15.201 Frequencies of operation.

(a) A low power communication device may be operated on any frequency in the bands 10–490 kHz, 510–1600 kHz and 26.97–27.27 MHz.

(b) Other frequencies above 70 MHz may be used for operations of short duration in accordance with the requirements set forth in § 15.211.

(c) Telemetering devices and wireless microphones may be operated in the band 88–108 MHz in accordance with the provisions of § 15.212.

(d) A low power communication device used for measurement of the characteristics of materials may be operated on frequencies and under the alternative provisions listed in § 15.214.

§ 15.202 Radiation limitation below 1600 kHz.

A low power communication device which operates on any frequency between 10 and 490 kHz or between 510 and 1600 kHz shall limit the radiation so that the field strength does not exceed the value specified in the following table:

Frequency kHz	Distance (feet)	Field strength ($\mu V/m$)
10–490	1,000	$\frac{2400}{F(kHz)}$
510–1600	100	$\frac{24000}{F(kHz)}$

§ 15.203 Alternative requirement for operation on frequencies between 160 and 190 kHz.

In lieu of meeting the radiation limitation, stated in § 15.202, a low power communication device operating on a frequency between 160 and 190 kHz need only meet the following requirements:

(a) The power input to the final radio frequency stage (exclusive of filament or heater power) does not exceed one watt.

(b) All emissions below 160 kHz or above 190 kHz are suppressed 20 db or more below the unmodulated carrier.

(c) The total length of the transmission line plus the antenna does not exceed 50 feet.

§ 15.204 Alternative requirement for operation on frequencies between 510 and 1600 kHz.

In lieu of meeting the radiation limitation stated in § 15.202, a low power communication device operating on a frequency between 510 and 1600 kHz inclusive need only meet the following requirements:

(a) The power input to the final radio stage (exclusive of filament or heater power) does not exceed 100 milliwatts.

(b) The emissions below 510 kHz or above 1600 kHz are suppressed 20 dB or more below the unmodulated carrier.

(c) The total length of the transmission line plus the antenna does not exceed 10 feet.

(d) Low power communication devices obtaining their power from the lines of public utility systems shall limit the radio frequency voltage appearing on each power line to 200 microvolts or less on any frequency from 510 kHz to 1600 kHz. Measurements shall be made from each power line to ground both with the equipment grounded and with the equipment ungrounded.

NOTE: One method of determining radio frequency voltage on the power line is described in "Military Specification for Interference Measurement" MIL-I-16910 (SHIPS) dated January 14, 1952, available from the Commanding Officer, Naval Supply Depot, Scotia, New York, 12302. Note that this procedure calls for grounding the equipment under test, whereas the Commission's rules call for measurements both with the equipment grounded and with the equipment ungrounded.

§ 15.205 Operation within the frequency band 26.97-27.27 MHz.

A low power communication device may operate within the band 26.97-27.27 MHz (27.12 MHz±150 kHz) provided it complies with all of the following requirements:

(a) The carrier of the device shall be maintained within the band 26.97-27.27 MHz.

(b) All emissions, including modulation products, below 26.97 MHz or above 27.27 MHz shall be suppressed 20 dB or more below the unmodulated carrier.

(c) The power input to the final radio stage (exclusive of filament or heater power) shall not exceed 100 milliwatts.

(d) The antenna shall consist of a single element that does not exceed 5 feet in length.

§ 15.211 Operation above 70 MHz.

(a) Except for telemetering devices and wireless microphones operated in accordance with §§ 15.212 and 15.213, and radio controls for door openers operating in accordance with § 15.215, a low power communication device manufactured on or after July 15, 1963, may be operated on frequencies above 70 MHz, provided it complies with all of the following conditions:

(1) The radiated field on any frequency from 70 MHz up to and including 1000 MHz does not exceed the limits specified for receivers in § 15.63.

(2) The radiated field on any frequency above 1000 MHz does not exceed 500 microvolts per meter at a distance of 100 feet.

(3) The device is provided with means for automatically limiting operation so that the duration of each transmission shall not be greater than 1 second and the silent period between transmissions shall not be less than 30 seconds.

(4) The device shall be so constructed that there are no external or readily accessible controls which may be adjusted to permit operation in a manner inconsistent with the provisions of this paragraph.

(b) Except for radio controls for door openers and for telemetering devices and wireless microphones operated in accordance with the requirements of §§ 15.212 and 15.213, a low power communications device, manufactured before July 15, 1963, may be operated on any frequency above 70 MHz: *Provided*, it complies with all of the following conditions:

(1) The radiated field on any frequency from 70 MHz up to and including 1000 MHz does not exceed the limits specified for receivers in § 15.63.

(2) The radiated field on any frequency above 1000 MHz does not exceed 500 microvolts per meter at a distance of 100 feet.

(3) The device is provided with means for automatically limiting operation to a duration of not more than 1 second, not to occur more than once in 30 seconds.

[§ 15.211(a) *amended, par (5) & (6) deleted eff. 5-7-71, this date later stayed to 11-1-71; II(69)-7*]

§ 15.212 Telemetering devices and wireless microphones in the band 88-108 MHz.

(a) Operation in the band 88-108 MHz is limited to low power communication devices employed as telemetering devices or as wireless microphones. This band shall not be used for two way communication.

(b) Users of these devices shall take adequate precautions to insure that harmful interference is not caused to the reception of transmissions from any FM or television broadcast station or any other class of station licensed by the Commission. In the event that such interference does occur, operation of the telemetering device or wireless microphone shall be promptly suspended and shall not be resumed until the interference has been eliminated. Users of these devices must accept any interference which may be caused by the operation of any licensed station operating in accordance with the terms of its license.

(c) Emissions from the device shall be confined within a band 200 kHz wide centered on the operating frequency. Such 200 kHz band shall lie wholly within the frequency range 88-108 MHz.

(d) The field strength of emissions radiated within the specified 200 kHz band shall not exceed 50 µV/m at a distance of 50 feet or more from the device.

(e) The field strength of emissions radiated on any frequency outside the specified 200 kHz band shall not exceed 40 µV/m at a distance of 10 feet or more from the device.

(f) Except as provided in § 15.213, no such device shall be operated unless it has been type approved pursuant to § 15.235.

(g) No antenna other than that furnished by the manufacturer shall be used with any type approved device.

§ 15.213 Custom built telemetering devices.

Custom built telemetering devices used for experimentation by an educational institution need not be type approved, *Provided*:

(a) The device complies with the technical requirements of § 15.212;

(b) The device has been certificated pursuant to §§ 15.227 and 15.228; and

(c) The educational institution notifies the Engineer in Charge of the local FCC office, in writing, in advance of operation. The notice shall include:

(1) The dates and place where the device will be operated;

(2) The purpose for which the device will be used;

(3) A description of the device including the operating frequency, RF power output, and antenna; and

(4) A statement certifying that the device complies with the technical provisions of § 15.212.

§ 15.214 Alternative provisions for measuring devices.

(a) A low power communication device used for measurement of the characteristics of materials may operate in the frequency bands listed in paragraph (c) pursuant to the provisions in this section.

(b) A device operated pursuant to the alternative provisions of this section may not be used for voice communications, or the transmission of any other type message.

(c) The device shall operate within the frequency bands:

MHz	MHz
13.554-13.566	890-940
26.96-27.28	(See note)
40.66-40.70	2400-2500
	5725-5875
	22000-22250

NOTE: The frequency band 890-940 MHz is subject to change pursuant to the reallocation of frequencies that may be made in the band 806-960 MHz in the rule making proceeding in Docket No. 18262.

(d) The maximum level of emission from the device shall not exceed:

Fundamental frequency in the band	Emission (μv/m at 100 feet)		
	On fundamental frequency	On harmonic frequencies	On other frequencies
13.554–13.566 MHz	15	0.5	0.5
26.96–27.28 MHz	32	1.0	1.0
40.66–40.70 MHz	50	1.5	1.5
above 890 MHz	500	50.0	15.0

(e) The device shall be self-contained with no external or readily accessible controls which may be adjusted to permit operation in a manner inconsistent with the provisions of this section. Any antenna that may be used with the device shall be permanently attached thereto and shall not be readily modifiable by the user.

(f) The device shall be prototype certificated pursuant to §§ 15.251–15.254 inclusive.

§ 15.215 Provisions for a radio control transmitter for a door opener.

(a) A low power communication device used for the radio control of a door opener may operate on any frequency above 70 MHz subject to the provisions of this section.

(b) The device may be used only for the purposes of opening or closing a door and may not be used for voice transmission or the transmission of any other type of message.

(c) Emission of RF energy from the transmitter, as well as from the receiver part of the control, shall not fall within any of the bands listed below:

MHz	MHz	GHz
73–75.4	608–614	10.68–10.70
108–118	960–1215	15.35–15.4
121.4–121.6	1400–1427	19.3–19.4
242.8–243.2	1535–1670	31.3–31.5
265–285	2690–2700	88–90
328.6–335.4	4200–4400	
404–406	4990–5250	

(d) Subject to the limitation in paragraph (c) of this section, emission of RF energy from the transmitter shall not exceed the levels given below when measured under open field conditions as prescribed in FCC Technical Division Report T–7001 dated October 1, 1970.

Frequency (MHz)	Field strength at 100 ft. (μV/m)
70–130	125.
130–174	125–375 (linear variation).
174–260	375.
260–470	375–1250 (linear variation).
Over 470	1250.

(e) The transmitter part of the control shall be activated only by a switch which will automatically deactivate the transmitter when released. The switch shall be of such quality to insure reliable operation for the expected life of the transmitter.

(f) The transmitter part of the radio control must be certificated pursuant to §§ 15.260–15.266.

(g) The receiver part of the radio control shall be separately certificated pursuant to Subpart C of this part.

[§ 15.215 added eff. 5–7–71, this date later stayed to 11–1–71; II(69)–7]

§ 15.220 Eavesdropping prohibited.

As provided in § 15.11, the use of a low power communication device for eavesdropping is prohibited.

§ 15.221 Class B emission prohibited.

Operation of low power communication devices that produce Class B emissions (damped waves) is prohibited.

§ 15.222 Interference from low power communication devices.

Notwithstanding the other requirements of this part, the operator of a low power communication device, regardless of date of manufacture, which causes harmful interference to an authorized radio service, shall promptly stop operating the device until the harmful interference has been eliminated.

§ 15.227 Certification requirements.

(a) Except for telemetering devices and wireless microphones which have been type approved pursuant to § 15.235, no low power communication device manufactured after the dates set forth in § 15.229 shall be operated without a station license unless it has been certificated to demonstrate compliance with the requirements in this part.

(b) The owner or operator need not certificate his own low power communication device, if it has been certificated by the manufacturer or distributor.

(c) Where certification is based on measurement of a prototype, a sufficient number of units shall be tested to assure that all production units comply with the technical requirements of this subpart.

(d) The certificate may be executed by a technician skilled in making and interpreting the measurements that are required to assure compliance with the requirements of this part.

(e) The certificate shall contain the following information:

(1) The operating conditions under which the device is intended to be used.

(2) The antenna to be used with the device.

(3) A statement certifying that the device can be expected to comply with the requirements of this subpart under the operating conditions specified in the certificate.

(4) The month and year in which the device was manufactured.

§ 15.228 Location of certificate.

The certificate shall be permanently attached to the device and shall be readily visible for inspection.

§ 15.229 Date when certification is required.

All low power communication devices which operate on frequencies of 70 MHz or above, manufactured after June 30, 1958, shall comply with the type approval or certification requirements of this subpart. All low power communication devices which operate on frequencies below 70 MHz, manufactured after December 31, 1957, shall comply with the certification requirements of this subpart.

§ 15.235 Type approval.

(a) A manufacturer of a telemetering device or wireless microphone who desires to obtain type approval for his equipment may request permission to submit such equipment to the Commission for testing by following the procedure set out in Subpart F of Part 2 of this chapter, as modified by this section. The manufacturer shall furnish the following with his request for type approval:

(1) A report of measurements showing that the equipment is capable of complying with the requirements of § 15.212;

(2) A statement that at least 10 units are proposed to be manufactured; and

(3) A statement agreeing to include a reprint of Subparts A and E of this Part 15, current as of date of manufacture, with each unit offered for sale or resale to the public.

(b) To receive type approval, telemetering devices and wireless microphones must meet the following requirements:

(1) The device must comply with the technical limitations of § 15.212.

(2) The design and construction of the equipment must give reasonable assurance of compliance with the

requirements of § 15.212 for at least 5 years under normal operation and with average maintenance.

(3) The device must be so constructed that the adjustment of any control accessible to the user shall not cause operation in violation of § 15.212.

§ 15.236 Identification of type approved devices.

The Commission will assign a type approval number to each telemetering device or wireless microphone which is type approved. The type approval number and the following statement shall be permanently inscribed upon or permanently attached to each production unit as follows:

> FCC Type Approval No. ----------
> Valid only when operated pursuant to FCC Rules, Part 15, and when used with antenna furnished by manufacturer

§ 15.237 Changes in type approved equipment.

No changes whatsoever may be made in a type approved telemetering device or wireless microphone, including the antenna, except on specific prior approval by the Commission.

§ 15.238 Withdrawal of certificate of type approval.

(a) A certificate of type approval may be withdrawn if the type of equipment for which it was issued proves defective in service and, under usual conditions of maintenance and operation, such equipment cannot be relied on to meet the conditions set forth in this part for the operation of the type of equipment involved, or if any change whatsoever is made in the construction of equipment sold under the certificate of type approval issued by the Commission, without the specific prior approval of the Commission.

(b) The procedure for withdrawal of the certificate of type approval shall be the same as that prescribed for revocation of a radio station license pursuant to the provisions of the Communications Act of 1934, as amended.

(c) In the case of withdrawal of a certificate of type approval, the manufacturer shall make no further sale of equipment under such certificate.

(d) When a certificate of type approval has been withdrawn for unauthorized changes or for failure to comply with technical requirements, the Commission will consider that fact in determining whether the manufacturer in question is eligible to receive any new certificate of type approval.

§ 15.251 Certification of measuring device operating pursuant to § 15.214.

(a) A device operating pursuant to § 15.214 need not be certificated by the owner or user if the device has been certificated by the manufacturer.

(b) Where certification is based on measurement of a prototype, a sufficient number of units shall be tested to insure that all production units can be reasonably expected to comply with the applicable technical requirements.

(c) The certificate shall be filed with the FCC, Washington, D.C. 20554.

(d) The certificate filed by the manufacturer will be available for public inspection pursuant to the provisions of §§ 0.457 and 0.461 of this chapter.

§ 15.252 Content of certificate required by § 15.214.

(a) The manufacturer, model, and serial number(s) or other positive identification of the device that was tested.

(b) Photographs of the device.

(c) A description of the circuitry and how the device operates.

(d) The conditions under which the device shall be operated.

(e) The antenna, if any, to be used with the device.

(f) A report of measurements pursuant to § 15.253.

(g) If filed by manufacturer, a statement certifying that production will be adequately controlled to insure that all units produced can be reasonably expected to comply with the applicable technical requirements.

(h) If filed by manufacturer, a copy of the installation and operating instructions provided to the user.

(i) Date of certificate.

(j) Signature. If filed by the manufacturer, the certificate shall be signed by a responsible official, who shall state that he is authorized to sign for the manufacturer and shall indicate his title.

§ 15.253 Report of measurements for a device operating pursuant to § 15.214.

The report of measurements may be prepared by any engineer skilled in making and interpreting the measurements that are required and shall contain the following information.

(a) Identification of the device(s) tested.

(b) List of measuring equipment used showing manufacturer, model number and date when last calibrated.

(c) Description of measurement procedure used. If a published standard was followed, reference to the standard is sufficient, provided any departure from such standard is described in detail.

(d) Report of the measurements obtained on the fundamental, and on harmonic and other spurious signals emitted by the device. For this measurement, the frequency spectrum shall be scanned from the lowest frequency generated by the device to the 10th harmonic of the operating frequency.

(e) Representative calculations used to determine field strength from the actual meter reading indicating the conversion factors used and their source.

(f) The date the measurements were made.

(g) The name and address of the engineer or technician who made the actual measurements, and the name and address of his employer, if any.

(h) The signature and printed name and address of the engineer responsible for the report.

§ 15.254 Identification of a device certificated under § 15.214.

(a) Each device certificated under § 15.214 shall be identified by a label which may be part of the nameplate.

(b) The label shall state that a certificate has been filed with the Commission attesting compliance with the applicable technical requirements.

(c) The label shall state further:

Operation of this equipment is subject to the following two conditions: 1. This equipment may not cause harmful interference. 2. This equipment must accept any interference that may be received, including interference that may cause undesired operation.

(d) The label shall be permanently attached to the device and shall be readily visible by prospective purchasers.

(e) The label may be attached only after the certificate required by § 15.214 has been filed with the Commission.

§ 15.260 Application for certification of a radio control transmitter for a door opener.

The application for certification of a radio control transmitter for a door opener shall contain the following information.

(a) The full name of the applicant.

(b) Mailing address.

(c) Name and title of responsible individual to be contacted for information concerning the application.

(d) Trade name(s) under which the device will be sold.

(e) Model number (or other positive identification) of each of the receiver and transmitter parts of the device.

(f) An expository statement describing how the device operates. This statement should include a block diagram, a circuit diagram, a description of the circuitry in the device, and a description of the antenna used with the device.

(g) Photographs of the device: Such photographs shall be 8″ x 10″, and shall clearly show the construction and circuit layout of the device. At least one exterior view shall be furnished showing the antenna and the controls available to the user. A sufficient number of views of the interior shall be furnished to define component placement and chassis assembly.

(h) A description of the installation and operating conditions that must be observed to insure the device will comply with the requirements of this part.

(i) A report of measurements pursuant to §§ 15.264-15.266.

(j) A copy of the installation and operating instructions furnished to the user. A draft copy of such instructions may be submitted with the application, provided a copy of the actual document to be furnished to the user is submitted within 45 days after certification is granted.

§ 15.262 Who may sign the application.

Each application, including amendments thereto, and related statements of fact required by the Commission, shall be personally signed by the applicant if the applicant is an individual; by one of the partners if the applicant is a partnership; by an officer if the applicant is a corporation; or by a member who is an officer if the applicant is an unincorporated association: *Provided, however,* That it will be sufficient if the application is signed by the head of an entity's engineering, technical, production, etc., department, with an indication of that representative's title, such as plant manager, etc.

§ 15.264 Report of measurements for certification of a radio control transmitter for a door opener.

The report of measurements shall be personally signed by the person who performed or supervised the tests who shall attest to the accuracy of the data and shall attach a brief statement of his qualifications. The report shall include the following information for both the transmitter and receiver parts of the device:

(a) Identification of the unit that was tested. Give name of manufacturer, model number, serial number.

(d) The latest calibration date of the measuring equipment and the standard against which such equipment was calibrated.

(e) A tabulation of the data that was obtained. The data shall show the actual meter reading, the computed value of the field strength, and shall include representative calculations showing how the field strength was determined from the meter source of such factor.

(f) A graphical display of the data. This may be a polar plot of field strength measured around the device.

(g) Date the measurements were made.

§ 15.266 Frequency range of transmitter measurements.

For devices operating below 500 MHz, the spectrum shall be scanned from the lowest frequency generated in the device up to 1000 MHz. For equipments operating on 500 MHz or higher, the spectrum shall be scanned from the lowest frequency generated in the device to 10 GHz. Measurements shall be made of all significant emissions observed. Emissions more than 20 dB below the permitted level need not be reported.

§ 15.268 Identification of a certificated radio control for a door opener.

(a) Irrespective of operating frequency, and notwithstanding the other labeling requirements of this chapter, each transmitter unit and each receiver unit of a radio control for a door opener shall be identified by an appropriate label containing the following information:

(1) The name of the grantee of certification or the trade name specified in the application for certification.

(2) The model number as given in the application for certification.

(3) The following statement:

This device complies with FCC Rules Part 15. Operation of this device is subject to the following two conditions: 1. This device may not cause harmful interference. 2. This device must accept any interference that may be received, including interference that may cause undesired operation.

(b) The information required by paragraph (a) of this section shall be permanently attached to the device and be readily visible to the prospective purchaser and user.

§ 15.270 Changes in a certificated radio control for a door opener.

No changes may be made in a certificated radio control for a door opener which alter the characteristics of the device that are required to be reported. If such changes are made, a new application for certification shall be filed.

§ 15.272 FCC Inspection.

Upon the request of the Commission, each grantee of certification shall be required, where practicable, to permit inspection of:

(a) Any device for which certification has been granted;

(b) The grantee's quality control procedures, inspection, and test data, and materials, and testing devices;

(c) The manufacturing plant and facilities; and

(d) The technical data files on that article.

§ 15.274 Interference from a radio control for a door opener.

(a) Operation of a radio control for a door opener is subject to the general conditions of operation set out in §§ 15.3 and 15.222.

(b) The operator of a radio control for a door opener who is advised that his device is causing harmful interference to an authorized radio service shall promptly stop operating the device, and he shall not resume operation until the condition causing the harmful interference has been eliminated.

[§§ 15.260-15.274 added eff. 5-7-71; § 15.268 amended and eff. date stayed to 11-1-71; 11 (69-7)]

§ 15.276 Date when certification of a radio control for a door opener is required.

(a) A radio control for a door opener that operates above 70 MHz marketed on or after November 1, 1971, shall comply with the certification requirements of § 15.215 and §§ 15.260-15.266 inclusive.

(b) No radio control for a door opener that operates above 70 MHz marketed prior to November 1, 1971, may operate after October 1, 1978, or such earlier date that may be specified by the Commission if the receiver thereof is found to be a source of harmful interference, unless the receiver has been certificated to demonstrate compliance with the provisions of § 15.63(d) and the transmitter has been certificated to demonstrate compliance with the provisions of § 15.215.

[§ 15.276 added eff. 11-1-71; 11 (69-7)]

Subpart F—Field Disturbance Sensors

[Subpart F (§§ 15.301-15.317) added eff. 10-5-71; 11 (69)-8]

§ 15.301 Scope of this subpart.

This subpart provides rules governing the operation of restricted radiation devices which are used as field disturbance sensors. Typical examples of devices regulated by these rules are microwave intrusion sensors and devices that use RF energy for production line counting and sensing.

§ 15.303 Restriction on operation.

No field disturbance sensor may be operated unless it has been certificated and labeled as complying with the requirements of this part.

§ 15.305 General technical specification.

(a) A field disturbance sensor may be operated on any frequency (including frequencies above 000 MHz) subject to the requirement that the field strength of emissions on the fundamental or on a harmonic or on other spurious frequencies shall not exceed 15 uV/m at a distance of $\lambda/2\pi$ from the sensor.

NOTE: The distance $\lambda/2\pi$ is equivalent in feet to 157 divided by the frequency in MHz.

(b) Alternative to paragraph (a) of this section, a field disturbance sensor may be operated on any frequency listed below, subject to the technical requirements set out in §§ 15.307 and 15.309 of this part.

915 MHz	10,525 MHz
2450 MHz	22,125 MHz
5800 MHz	

§ 15.307 Permitted bands of operation.

The carrier frequency of a field disturbance sensor operating on one of the frequencies listed in § 15.305(b) and any modulation components thereof shall be kept within the following band limits:

Nominal operating frequency (MHz)	Band Limits (MHz)
915	±13
2450	±15
5800	±15
10,525	±25
22,125	±50

NOTE: To minimize the possibility of out-of-band operation because of frequency drift due to aging of components or other causes, it is recommended that the carrier frequency be kept within the central 80 percent of the permitted band.

§ 15.309 Emission limitations.

(a) For a field disturbance sensor operating within any frequency band listed in § 15.307, the field strength of emissions on the fundamental shall be limited in accordance with the following:

Frequency (MHz)	Field Strength
915	
2450	50,000 uV/m at 100 ft.
5800	
10,525	250,000 uV/m at 100 ft.
22,125	

(b) Spurious emissions (including emissions on a harmonic of any frequency listed in paragraph (a) of this section) shall be suppressed at least 50 dB below the level of the fundamental; however, suppression below 15 uV/m at 100 ft. is not required.

NOTE: For pulsed operation, measured field strength shall be determined from the averaged absolute voltage during a 0.1 second interval when field strength is at its maximum value. Below 1000 MHz, the measurement bandwidth shall comply with the requirements set out in the American National Standards Institute Specifications C63.2–1963 and C63.3–1964. Above 1000 MHz the measurement bandwidth shall be 5 MHz.

§ 15.311 Interference from a field disturbance sensor.

(a) Operation of a field disturbance sensor is subject to the general conditions of operation set out in § 15.3.

(b) The operator of a field disturbance sensor who is advised that his sensor is causing interference to an authorized radio service shall promptly stop operating the sensor, and operation shall not be resumed until the condition causing the harmful interference has been eliminated.

§ 15.313 Certification of a field disturbance sensor.

The procedure for certification of a field disturbance sensor is identical to the procedure for a radio control for a door opener as set out in §§ 15.260–15.264 inclusive, and §§ 15.268–15.272 inclusive, except that § 15.264(b) shall not apply.

§ 15.315 Description of measurement procedure.

The report of measurements shall describe in detail the measurement procedure that was used. If a published standard was used, reference to the standard is sufficient, provided any departure from the standard is described in detail.

§ 15.317 Frequency range over which measurements are required.

(a) For a field disturbance sensor operating below 100 MHz, the spectrum shall be scanned from the lowest frequency generated in the device up to 1000 MHz. Field strength for all significant emissions shall be measured and reported.

(b) For a field disturbance sensor operating above 100 MHz the spectrum shall be scanned from the lowest frequency generated in the device up to 10 GHz, provided that for sensors operating on frequencies above 5 GHz, the spectrum shall be scanned to the highest frequency feasible, above 10 GHz. Field strengths of all significant emissions shall be measured and reported.

[Subpart F (§§ 15.301–15.317) added eff. 10–5–71; II(69)-8]

§ 2.701 Prohibition against use of a radio device for eavesdropping.

(a) No person shall use, either directly or indirectly, a device required to be licensed by section 301 of the Communications Act of 1934, as amended, for the purpose of overhearing or recording the private conversations of others unless such use is authorized by all of the parties engaging in the conversation.

(b) Paragraph (a) of this section shall not apply to operations of any law enforcement officers conducted under lawful authority.

Chapter 5

All About CB Antennas

Next to the transceiver itself, a proper antenna is the most important part of any two-way radio system. Selecting the right antenna for any particular installation involves many factors. None are especially difficult to understand, but all must be considered to obtain the best possible performance. Because the FCC limits the power of Citizens' Band transmitters to five watts, a CB user must make the best antenna choice and installation he can if he expects to get maximum range from his equipment.

The most common base station antenna is the "ground plane" type, so called because the three or four horizontal elements tend to direct radiated signals along the surface of the earth. Without these horizontal elements, the antenna would allow many radio waves to scatter toward the sky. Such "sky waves" are wasted for normal point-to-point communications as explained later in this chapter. When longer ranges between base stations are required, a "beam" type antenna may be used. These antennas look like big brothers of the type used for TV reception. Their elements direct and reflect the radio wave into a beam, so that the difference between this kind of antenna and a ground plane is much the same as the difference between an auto headlight and a bare bulb. Naturally, a beam won't give the broad blanket of coverage possible with a ground plane, but it is more effective for radio communication between two distant points; for instance, two farms about 25 miles apart.

Automobile CB antennas vary from the large 108-inch steel whip to the new compact types less than two feet long. The

94

This unusual antenna installation problem was solved by Jack Michols, KHH-4397, by using an extenson on the back of the tractor.

most effective mobile antenna would probably be the longer 108-inch whip type mounted in the middle of a sedan roof with the roof acting as a ground plane. However, low bridges and general over all appearance usually rule out this ideal installation. So the best practical antenna is the 108-inch whip on the left rear bumper. This permits installation without marring the auto body, and since the whole antenna radiates well, the best range will result. Where appearance is of prime importance, such as on the family car, coil-loaded antennas should be considered. While they are not as effective as a full-sized whip, their short length makes them virtually invisible. They may be mounted in the middle of a car roof like a taxi antenna, or attached in place of the normal car radio antenna.

Before purchasing an antenna, remember its importance in a two-way radio system. It will be exposed to wind, rain, ice, and heat, and will be difficult to maintain after installation. The difference in price between a good antenna and a budget one is much less than a single antenna service bill.

BEAM ANTENNAS

A beam antenna is used to provide communications between

Basic 108-inch whip antenna. This whip has ⅜–24 threads that mate with the familiar ham-type mobile spring mount. (The spring mount itself requires that 3 small holes and 1 large one be drilled in the car body.)

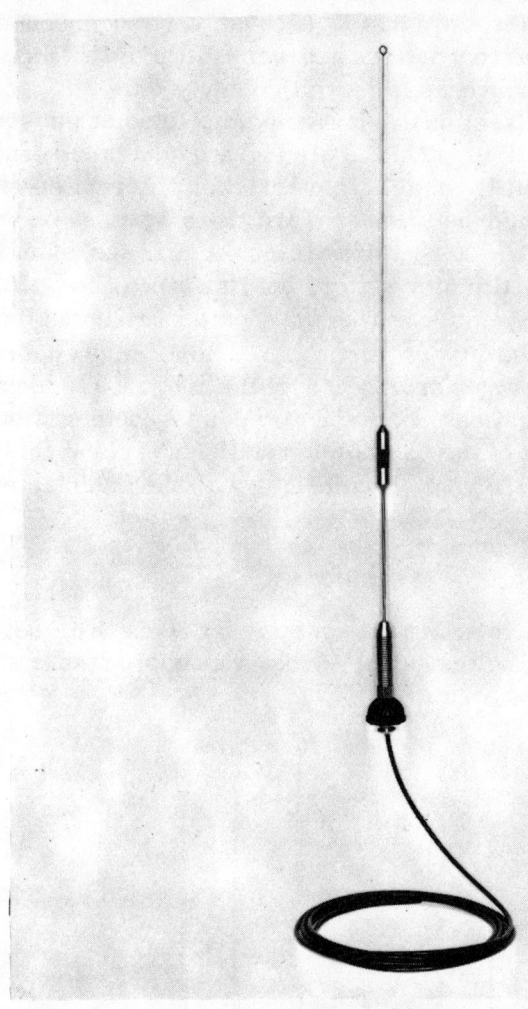

Centerloaded CB whips, often furnished complete with cable system (like the one above), are available at low cost from most CB distributors and dealers. Notice spring mount, which will "give" in high wind.

two fixed stations. As the name says, this type of antenna directs the transmitted signal in a kind of "beam" toward the station to which it is pointed. Besides being directional, a beam antenna also has a given amount of gain which is a definite advantage as it increases the effective radiated signal. One of the reasons it does this is because a beam squeezes the signal into a narrow pattern and very little of the signal is sent up into space where it doesn't do any good.

Since a beam is directional, it has to be pointed at the station you want to talk to. This is all right if you talk to only that one station. But if you talk to others in a different direction, the beam should be turned toward those stations or the signals will be weak. A rotator will let you turn the beam in any direction at the flip of a switch, but this type of operation is a little more expensive. The wave pattern radiated by a beam antenna may be either vertically or horizontally polarized, depending on the orientation of the antenna elements. For best communications between two points, both stations should use antennas of the same polarization. Vertical polarization is preferred, since it coincides with the polarization of mobile antennas.

COAXIAL ANTENNAS

A coaxial antenna is a vertical system that takes little room to mount. One of the things that make a coaxial antenna ap-

Two 3-element vertical beams mounted for base station use. In practice, a rotator is used to direct the array to the station being "worked."

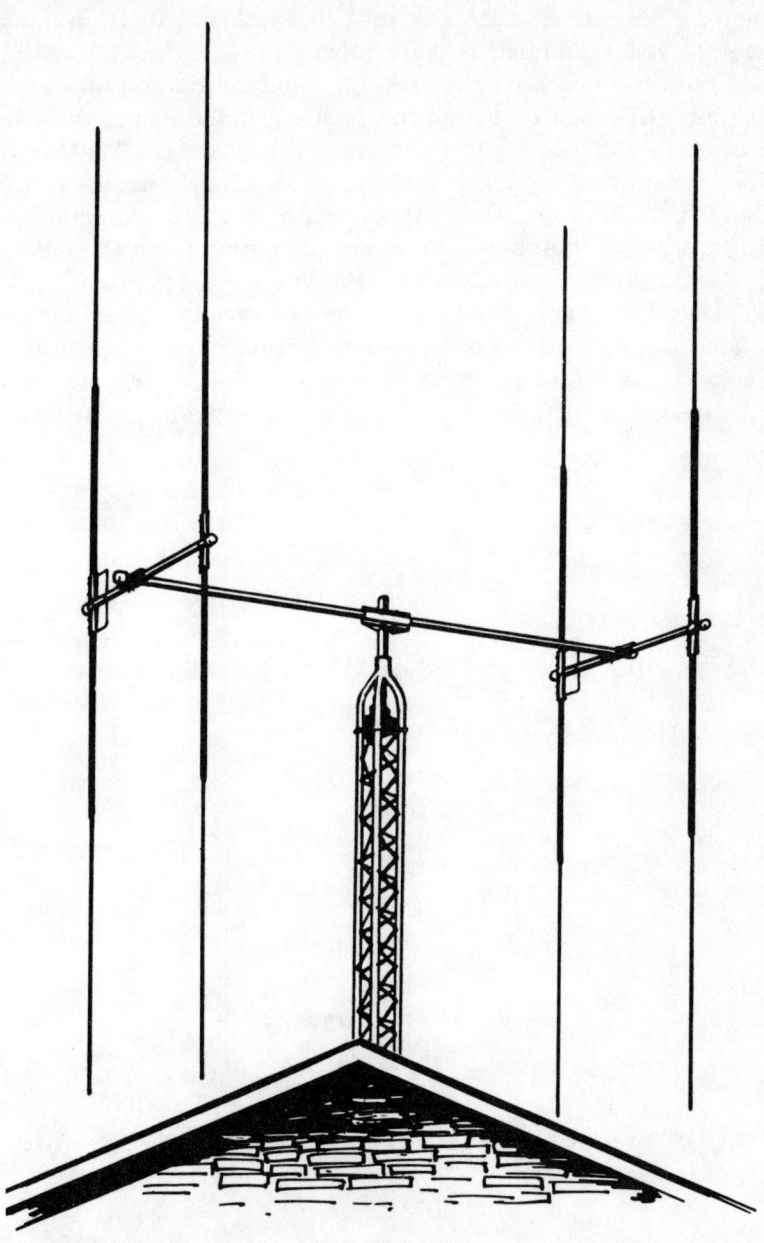

A rooftop "beam" antenna is actually two 2-element yagi antennas phased for optimum efficiency and mounted for vertical compatibility with mobile antennas.

pealing for CB operation is its low angle of radiation. This means that the signal is kept closer to the ground level and not radiated into space uselessly. Another reason for its use is that it is ideally suited for base-to-mobile operation because the radiation pattern is omni-directional. That means the transmitted signal is sent out in all directions. For most base-to-mobile operations this is the only way to communicate with vehicles which can be in any direction from the station. A coaxial antenna is actually a dipole mounted vertically, and it is vertically polarized as are the antennas used on vehicles.

The coaxial consists of two quarter-wave elements, each 108 inches long. One element is a whip, the other is a metal sleeve or "skirt" connected together through an insulating ma-

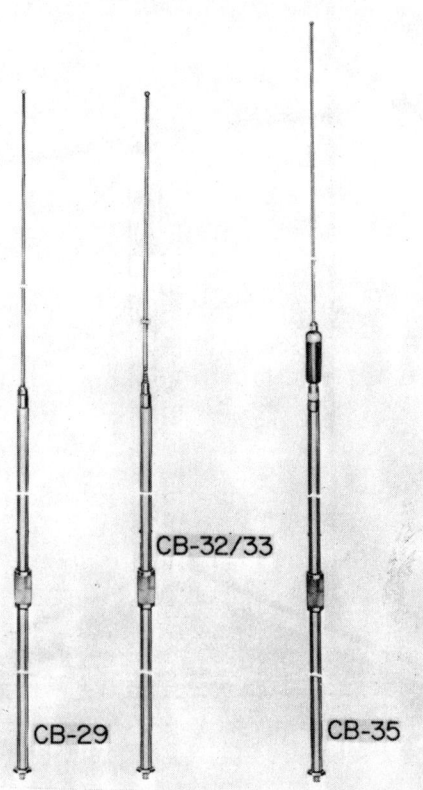

Shown here are three types of coaxial CB antennas, all of which can be mounted either on a rooftop or on a vehicle.

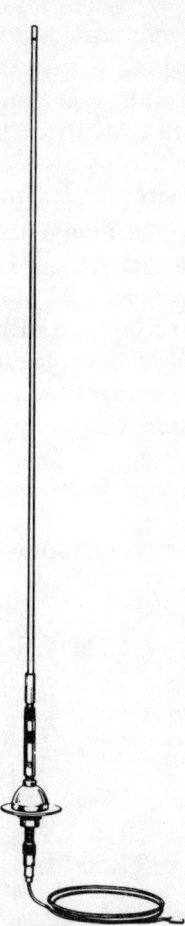

Vertical whip antenna, used commonly on cars equipped with CB gear. Prices are more than reasonable, and the antenna performs exceptionally well, even at high vehicular speeds.

terial at the center. The lead-in cable is normally fed up the center of this skirt and connects to the two elements through a coaxial fitting. The lead-in cable between the antenna and the transceiver is normally coaxial cable. The antenna is mounted on a pipe which goes up the center of the skirt and fits into the insulating material at the center of the two radiating elements. There also are insulating rings spaced along the inside of the skirt to prevent it from touching the mounting pipe.

The main advantage of using a coaxial antenna is that it takes very little space to mount, provides a low angle of radiation, it's omni-directional and vertically polarized, as are those used for mobile operation. It requires 72-ohm coaxial cable (RG-59/U), which is smaller is diameter and lighter than its 50-ohm equivalent (RG-58A/U). The disadvantages are the weight caused by the long pipe necessary for mounting and the fact that the coaxial-cable impedance of 72 ohms is not as common—or available—as conventional 50-ohm types. Also, the coaxial antenna has a high degree of frequency sensitivity. This means that the antenna will perform well at its resonant frequency only, and drop off on either side of this. In other words, it may be tuned to operate on Channel 15, and operation on other channel becomes less efficient as you get further away from this channel.

GROUND PLANE ANTENNA

A ground plane antenna is somewhat like a coaxial antenna

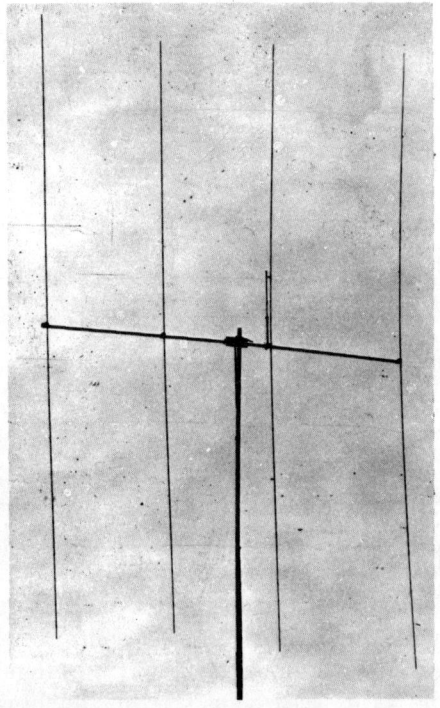

Here is a 4-element yagi beam antenna, quite popular these days for base station istallations. At far right is reflector element; next from right is the "driven" element (electrically much the same as the half-wave dipole). Two elements at left are directors.

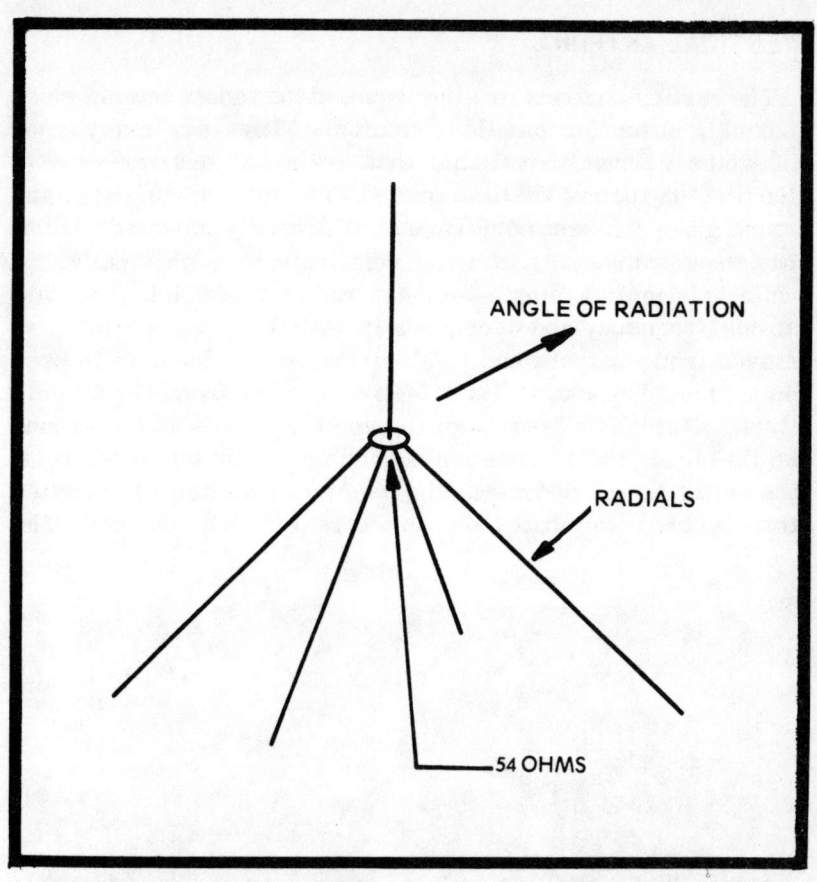

The "drooping" groundplane has an impedance of about 54 ohms and an angle of radiation of approximately 25 degrees. This type of antenna is frequently used for VHF work because of its omnidirectional character.

in that one of the radiating elements is a vertical whip. But instead of a sleeve or skirt for the other element, it has several quarter-wave arms or radials which provide an artificial ground. There are normally four radials extending out from the center point of the antenna and isolated from the vertical whip. This is one of the most popular antennas used for CB operation since it provides a vertically-polarized omnidirectional pattern. One of the advantages of this antenna is that it does not have to have a good physical ground to be effective since the radials serve as an artificial ground. However, it should be mounted at least a quarter-wave length above the ground level.

VERTICAL ANTENNA

The vertical antenna or whip is one of the most common types normally used for mobile operation. There are many types of whips currently available that are a full quarter-wave in length (108 inches) or base loaded. The full quarter-wave antenna gives the best performance if properly mounted, but it has the disadvantage of being relatively long physically. A whip is simply a single-element radiator which is resonant at one frequency and decreases in efficiency as operation is moved from that channel. A vertical whip, such as is used on a vehicle, obtains its effective ground from the vehicle itself. Actually, the roof of the vehicle serves as the ground so the closer the antenna can be mounted to the top of the roof, the better it will operate. The common practice of mounting these whips is to locate them on the rear deck or bumper. The

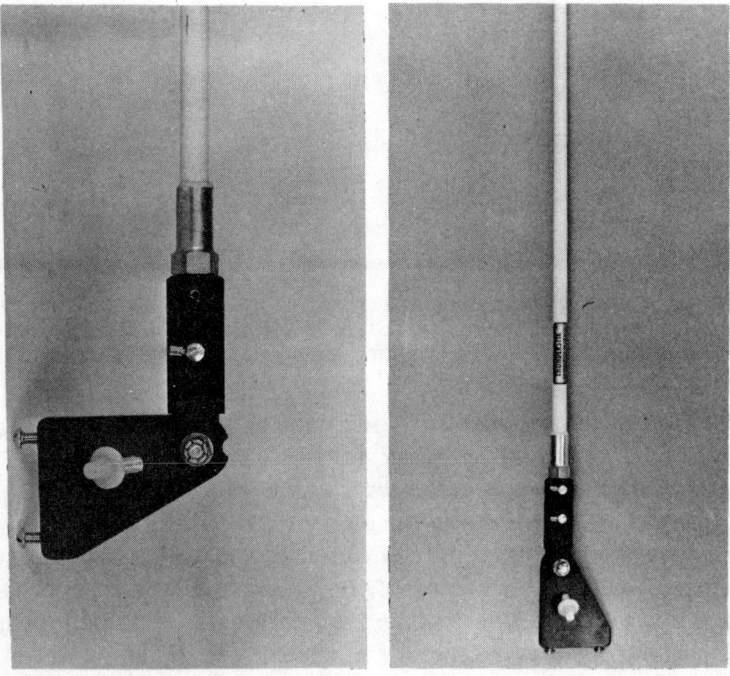

This CB antenna, called the Thunderstick, is designed to fold over when not in use. It mounts above the passenger's door, fitting neatly into the "rain gutter."

The antenna against which all others are referenced is the half-wave dipole. In photo above, engineers for Pearce-Simpson check transceiver efficiency with horizontally mounted dipole antennas, cut for the precise frequency of operation on the 27 MHz CB band.

rear deck is better than the bumper as far as the antenna is concerned, but it is somewhat unsightly to have 108 inches of whip sticking up that way, too. To get around this, many antenna manufacturers now have the base-loaded type whip which is physically very short, but electrically the same length as a quarter-wave. The base-loaded antenna has a disadvantage which may or may not be a factor in all cases, but it is worth mentioning. A base-loaded antenna uses a coil at the bottom of the antenna to electrically increase the length of the antenna. This means that some of the signal is going to be absorbed

105

by the coil and less of the signal will reach the vertical radiator. However, in most instances the loss of efficiency will not be too much of a factor.

A whip is omni-directional and when mounted on the rear deck of a vehicle the pattern is more directional over the roof toward the opposite corner from which it is mounted. If operation is to be on one channel, a whip can be tuned by increasing or decreasing its length. It is almost impossible to add to the length of the whip without adding a loading coil, so if you intend to trim it for resonance on one channel, be sure you purchase one longer than necessary to begin with. Tuning of any antenna should be done by cutting the length a little at a time and checking for a low VSWR (voltage standing wave ratio). More on this later.

Yes, they even make CB antennas suitable for mounting atop aircraft.

YAGIS AND QUADS

A yagi antenna is one that has a driven element, a parasitic reflector, and one or more directors. It is a "unidirectional" antenna which means that all the power is sent out in one direction. This type of antenna has a given amount of gain and can be tuned to provide a maximum front-to-back ratio. Front-to-back ratio is a term meaning that signals from any direction other than the front of the antenna will be suppressed or reduced. This is to prevent signals you are not pointed at from interfering. As indicated, a yagi can have more than one parasitic element, but only one reflector. The reflector is placed behind the driven element and the directors in front of it. The most popular yagi can simply be described as one having a half-wave dipole for the driven element with one director and one reflector.

The spacing between the parasitic elements and the driven element in a yagi is somewhat critical as it affects both frequency sensitivity and the front-to-back ratio. If a yagi is to be used over a band of frequencies, the elements are normally spaced further apart, and they are cut to progressively shorter length (relative to distance from driven element). This means that you sacrifice some gain, but overall performance across the band is better. The ideal use of a yagi would be to have it tuned for maximum gain and front-to-back ratio on one desired channel.

STACKED YAGIS

Yagi antennas can be combined or "stacked" to increase gain and directivity. The effect of stacking can increase your power as much as two times over the normal gain of one yagi or beam. They can be stacked by placing them one over the other or side by side. Spacing in these installations is critical as it affects both gain and directivity, but can be worthwhile if properly done. Your local electronic dealer can probably supply you with the necessary wiring to connect the lead-in to the antennas. But be sure to do it right or the cost will outweigh the advantages. Regardless of the antenna you buy, be sure to pick one designed to do the best job for your needs and above all, install it properly.

STANDING WAVES—WHAT THEY MEAN

Standing waves, better known in communications circles as

VSWR (voltage standing wave ratio) or SWR (standing wave ratio). Whichever you choose to call it, it is the loss resulting from a mismatched antenna system and the instrument used to measure it is called a VSWR (or SWR) bridge.

The CB transceiver output is made to match a 50-ohm antenna and the feed-in cable between the transmitter and antenna has to have the same value—50 ohms. If the antenna or lead-in cable looks back at the transmitter with some value other than 50 ohms, there is a mismatch and your transmitter power does not all get to the antenna. In fact, the transmitted power goes out to the antenna and some of it is "reflected" back to the transmitter. It is this "reflected" power that we call VSWR.

The amount of VSWR permissible in a CB communications system should be kept as low as possible because you have only a few watts of power and you don't want any wasted. A VSWR meter can be used to check your antenna system by inserting it in the lead-in close to the transmitter. On most VSWR meters there is a switch labeled "Forward" and "Reflected" and a variable control to set the full-scale reading. With the transmitter connected through the meter to the lead-in and antenna, you transmit and adjust the meter to read full-scale in the FORWARD position. Then switch to the REFLECTED position and read the amount of VSWR from the meter face. For example, suppose you key the transmitter and set the meter to read full-scale which may be 100% output. Then you switch to reflected and the meter drops to 30% —30% of your total power output is being reflected. That means you have a mismatch allowing only 70% of your power to reach the antenna. If your transmitter output is 3 watts, you are loosing almost one watt because of the mismatch!

This can mean that the lead-in cable is defective, or the antenna has some broken, bent, or missing elements. One way to find out is to connect a 50-ohm dummy load in place of the antenna—at the end of the lead-in. This will quickly tell you where the problem is. But remember, VSWR is caused by a mismatch in the lead-in or antenna, not by transmitter tuning. The transmitter can be tuned for maximum output, but not for minimum VSWR.

MAXIMUM RANGE

There are no cut-and-dried figures that can be given to any

one type of system that say you can operate 10, 15, or 20 miles. There are too many variable factors involved such as terrain, operating frequency, type of antenna, antenna height, power, and interference. Generally speaking, with a properly installed base station, you can communicate 10 to 15 miles to a mobile station. The limitations here are based more on the fact that a mobile station has a relatively low antenna, it is subject to ignition noise, voltage fluctuation, and building obstructions. A farmer in open country will do better range wise base-to-mobile than a business man using the same system in a city simply because he is not likely to have all the interference. Normally, operation over water will provide better operating distances. The operating distances between mobiles is considerably less than base-to-mobile because of antenna height and interference. Normal ranges vary from 2 to 7 miles depending upon conditions.

ANTENNA LIMITATIONS

As we indicated earlier, the antenna height is very important to good range. The FCC says that a CB antenna for Class D operation can be no more than 20 feet above any natural formation or man-made structure on which it is mounted, although it can be mounted on a ground-based tower so that the highest point of the antenna is no more than 60 feet above ground. There are other limits indicated in Part 95.37 of the FCC Rules and Regulations that deal with antennas mounted on other existing towers.

These limits say that you can mount your CB antenna on a tower used for another transmitting service as long as your antenna does not go more than 60-ft above ground. For example, if you have a 200-foot tower near you that is used for TV or for industrial communications, you can put your antenna on it as long as it doesn't go higher than 60 feet. If you have a situation where you can legally make use of these limits and greatly increase your antenna height, it may be worth looking into. Doubling the height of your antenna can be as effective as doubling the power of your transmitter.

However, you should be aware that the lead-in wire (generally coaxial cable) also has some losses and these losses increase with the length of the lead-in. If you have to run hundreds of feet of coax just to gain a few feet of antenna height, it may not be a gain at all. On the other hand, the new low-loss cables make it possible to use longer runs with very little loss. (See cables and losses in Chapter 6.)

RADIO WAVES

A radio wave is itself very complicated and it is not our intent to get into the highly technical theory of what makes a radio wave. To simply describe the CB radio wave for our purposes here, consider it as an unseen motion or "disturbance" created by energy sent to the antenna from our CB transmitter. This radio wave or motion goes out from the antenna over the earth's surface. If the antenna is omnidirectional, the waves go in all directions. If the antenna is a beam type, the waves are directional as indicated earlier. Some of these waves follow the curvature or surface of the earth—these are called "groundwaves," and the only type considered in CB radio transmission. (Other waves travel up into the atmosphere and are called "skywaves." It is these waves that provide the phenomenon known as "skip.") When these waves hit a receiving antenna they cause an electrical current to flow which goes down the lead-in and into the receiver where it is changed into an understandable message.

While "groundwaves" tend to follow the curvature of the earth, the distance covered depends on the wavelength or frequency of the signal. Low frequencies, such as those used in the standard broadcast radio band have very long wavelengths. It is not uncommon to hear a low-power BC station 100 miles away. The reason is because the earth absorbs less energy from these radio waves than from the higher frequencies used in CB radio. The higher the frequency, the more energy is absorbed. Since CB radio is thousands of cycles higher in frequency than the broadcast band, much more is absorbed and the shorter the range on groundwave communications. This also is the reason for using an antenna which directs most of its signal over the earth's surface and less toward the atmosphere.

POLARIZATION

We mentioned earlier that radio waves can be either vertically or horizontally polarized. It has been established also that all of the antennas in a particular communication system should have the same polarization. In other words, if the mobiles are vertically polarized, the base station should be too. Consider a whip antenna on a vehicle. When the radio waves leave the antenna, they set up a motion or disturbance which is vertical in reference to the earth's surface. When they hit a vertically-polarized antenna, they are able to set up a maximum amount of signal current flow. However, if the receiving antenna is horizontally polarized, much of the signal goes right on by with very little effect on the vertical antenna.

CITIZENS BAND ANTENNA DIRECTORY

Antennas for the 27-MHz CB band come in a wide variety of configurations and prices. The listing that follows is a result of a questionnaire mailed to known CB antenna manufacturers; naturally, some omissions will be noted. In all cases, statistics are those contributed by the antenna producer and do not represent results of tests by this publisher.

The first significant category in the directory is "type." Here again, the listing is that supplied by the manufacturer. Many sharply resemble conventional designs, such as coaxial, half-wave whip, yagi, etc. Generally, however, it can be assumed that "vertical" in the mobile category translates into a whip configuration; the "length" listing provides insight into wavelength fractions or possible loading coil construction. For the most part the gain, shown here in dB, is referenced over a typical half-wave 27-MHz dipole. This depends, however, on how individual antenna manufacturers measure "gain." Most seem to abide by the dipole reference, however.

Most antennas listed accept standard 50-ohm coaxial cable; many, indeed, are supplied with an appropriate length of RG-58A/U already terminated to the antenna and equipped with an Amphenol 83-1SP connector. Others come with a coaxial fitting for a cable connector plug. Still others, notably a few of the yagi/beam configurations, require that impedance-matching baluns be constructed by the installer. VSWR ratings, also supplied by manufacturer, are generally referenced to the bandwidth of the 27-MHz CB operating spectrum. As a result, they can be improved upon in many cases by trimming (adjusting T-matches, etc.) to the most-used operational frequency.

BASE ANTENNAS BOTH NEW AND "USED"

Mfg.	Model	Type	Gain (db)	VSWR	Length	Wgt. (lbs.)
Antenna Specialists	M-113	hor/vert	hor: 8 vert: 9.75	1.5:1	108" boom	26
	M-117	3-element half-wave shunt-fed	3.75	1.5:1	216"	10
	M-119	sector phased beam	7.75	1.5:1	17.5'	17
	M-134	vert/hor 5-element	hor: 11 vert: 12.5	1.5:1	22' boom	19
Apelco	BCL-1	vertical	–	–	234"	9
Avanti	Astro Plane	vertical	4	1.4:1	12'	3.5
	PLD	vertical	11	1.2:1	11.9'	13.5
Cush Craft	Ringo	vertical	3.75	1:1	214"	5
	CB-11	3-element beam	7.5	1:1	120"	15
	CB-114	4-element beam	9	1:1	192"	20
	CB-115	5-element beam	10	1:1	288"	25

Mfg.	Model	Type	Gain (db)	VSWR	Length	Wgt. (lbs.)
Cush Craft	CB-11D	6-element dual beam	10.5	1.2:1	216"	30
	CB-1140	8-element dual beam	12	1.2:1	216"	45
	TS-1	Universal	0	1.5:1	216"	2
	DGPA	Ground plane	0	1.2:1	216"	5
Francis Industries	Octopus	vertical	–	–	17'3"	–
Gam	Projector	vertical	1.5	1.5:1	18'	–
Hy-Gain	SDB4	vertical	9.2	1.1:1	9' cross 3'1" beam boom	16
	SDB6	vertical	12.7	1.1:1	14' cross 12'2" beam boom	30
	SDB10	vertical	13.9	1.1:1	24' cross 20' beam boom	55
	CB3	vertical	8	1.4:1	12' boom	7
	CB5	vertical	9.5	1.1:1	18' boom	18
	CBGP	vertical	unity	1.1:1	9' ht	3

	Model	Type	Gain	SWR	Length	Elements
	CBV	vertical	3	1.1:1	17' ht	6
	CLR2	vertical	3.4	1.1:1	19'10" ht	8
	GCLR2	vertical	3.4	1.1:1	19'10" ht	14
International	CBA-1	attic	—	2.1:0	18"	3
Crystal	CBD-1	dipole	—	2.1:0	36"	6
	CBG-1	vertical	—	2.1:0	18"	3
	CBB-1	ground plane	—	2.1:0	18"	3
	160-133	vertical	—	2.1:0	16"	10
Allied/Radio Shack	KN-2574	beam	12.3	1.5:1	18'8– 3/4" each element	15
	KN-2505	vertical	unity	1.5:1	9'	4
	KN-2570	vertical	5	1.5:1	9'	4
	KN-2572	vertical	6	1.2:1	19'8"	7
Lafayette	Range Boost II	vertical	3.75	1.17:1	210"	10
	5-element beam	vertical	10	1.5:1	204" boom	15
	3-element beam	vertical	8	1.6:1	198" elements	11
	4-element GP	vertical	—	1.8:1	108" element 108" radials	4

Mfg.	Model	Type	Gain (db)	VSWR	Length	Wgt. (lbs.)
Mark Products	MK-11	vertical	–	–	19'	12
	MJ-27	vertical	–	–	19'	11
	MJ-3	beam	–	–	19'	23
	CBB-1	vertical	–	–	17'	8
	MK-V	vertical	–	–	20'	23
Master Mobile	GPC	vertical	3.7	1.5:1	17'3"	–
Mosley Electronics Inc.	X-27-3	beam	8	1.5:1	224-3/4"	12.5
	X-27-4	beam	8.7	1.5:1	224-3/4"	15
	X-27-5	beam	9.5	1.5:1	224-3/4"	16.5
	AD-311	beam	8	1.5:1	216-1/2"	14
	A-311-S	beam	8	1.5:1	224-3/4"	12.5
	A-411-S	beam	8.7	1.5:1	224-3/4"	15
	SA-511-S	beam	9.5	1.5:1	224-3/4"	20.5
	SKT-3	beam and stacking kit	11	1.5:1	224-3/4"	29
	SKT-4	beam and stacking kit	12	1.5:1	224-3/4"	38
	SKT-5	stacking kit	13	1.5:1	224-3/4"	47

	A-311-511-S	conversion kit	–	–	10
	A-311-SK	stacking kit	–	–	18
	H-CB-SK	stacking kit	–	–	8
	Devant 1	vertical	1.5:1	235.5"	7.5
	Devant Special	vertical	1.5:1	245"	7.6
New-Tronics Corp.	PRO-27-SD	vertical	–	19'8"	16
	PRO-27	vertical	–	19'8"	15
	PRO-27-JR	vertical	–	19'10–3/4"	8.75
	GP-1	vertical	–	–	6
	11M-3	beam	–	18'10" boom	8.5
	11M-4	beam	–	14' 1" boom	10.75
	11M-44	stack 4 beams	–	–	–
	RTG-27-L	gutter lamp	–	25"	–
Polygon	CBQ-2	vertical	1.1:1	–	11
	CBQ-3	vertical	1.1:1	–	16
	CBQ-4	vertical	1.1:1	–	22

Mfg.	Model	Type	Gain (db)	VSWR	Length	Wgt. (lbs.)
Shakespear (C/P Corp.)	176	vertical	–	1.5:1	18' 6"	7.5
Webster	BCL-1	vertical	3.4	1.1:1	234"	9
	BCX-1	vertical	unity	1.4:1	216"	7

MOBILE ANTENNAS

Mfg.	Model	Type	Gain (db)	VSWR	Length	Wgt. (lbs.)
Antenna Specialists	M-1	whip, base spring	–	1.5:1	108"	6
	MR-58	trunk mount	–	1.5:1	48"	1.5
	M-67	roof mount	–	1.5:1	44"	1.5
	M-73	trunk mount	–	1.5:1	48"	2
	M-74	fender mount	–	1.5:1	44"	1.5
	M-103	combination CB/AM	–	1.5:1	46"	1.5
	M-123	roof mount	–	1.5:1	44"	1.5
	M-124	roof mount	–	1.5:1	46"	1.5
	M-125	roof mount	–	1.5:1	46"	1.5
	M-130	roof mount	–	1.5:1	18"	1.25
	M-131	gutter mount	–	1.5:1	18"	1.25
	M-168	magnetic mount	–	1.5:1	–	1.25
	M-176	trunk mount	–	1.5:1	46"	1.5

	M-180	gutter mount	–	–	–
Apelco	A-48	vertical	–	–	48"	1.5
	A-49	vertical	–	1.5:1	96"	1.5
	A-50	vertical	–	–	49"	1.5
	A-85	CB/AM	–	–	103"	1.5
Francis Industries	CB-14	vertical	–	–	48"	2
	Monowhip		–	–	96"	–
	CB-22 Tweeter	vertical	–	–	30"	–
	CB-24 Woofer	vertical	–	–	48"	–
	CB-26 Hot Rod	vertical	–	–	66"	–
	CB-28 Wheeler Dealer	vertical	–	–	96"	–
	CB-50 Amazer	vertical	–	–	96"	–
	CB-229 Tweeter	vertical	–	–	30"	–
	CB-249 W/PL-259	vertical	–	–	48"	–
Gam	8-Ball	vertical	1.5	1.5:1	86"	–

Mfg.	Model	Type	Gain (db)	VSWR	Length	Wgt. (lbs.)
General Radiotelephone Co.	M-27	vertical	–	1.2:1	–	–
Hy-Gain	CBW	vertical	unity	1.1:1	26"	1
	TJCQ	vertical	unity	1.1:1	46"	2
	TMCQ	vertical	unity	1.1:1	50"	2
	TMPQ	vertical	unity	1.1:1	46"	2
	TQCW	vertical	unity	1.1:1	32"	1
	TQMA	vertical	unity	1.1:1	59"	3
	TQRDX	vertical	unity	1.1:1	50"	2
	TQRMB	vertical	unity	1.1:1	28"	2
	TQW	vertical	unity	1.1:1	46"	1
	TRQS	vertical	unity	1.1:1	19"	2
	TTMPQ	vertical	unity	1.1:1	46"	3
	W102	vertical	unity	1.1:1	102"	3
	Hellcat One	vertical	unity	1.1:1	47"	3
	Hellcat Two	vertical	unity	1.1:1	23 7/8"	1.7
	Hellcat Three	vertical	unity	1.1:1	34 5/8"	3.2
	Hellcat Four	vertical	unity	1.1:1	47"	3.1

Manufacturer	Model	Type		SWR	Length	
International Crystal	CBM-1	vertical	–	2.1:0	18"	3
	160-134	vertical	–	2.1:0	30"	3
Allied/Radio Shack	17-6756	vertical	–	–	30"	1
	17-6757	AM–FM/CB	–	–	50"	1
	17-6705	vertical	–	–	102"	7
Lafayette	Base Loaded "Auto Top" Mount Antenna	vertical	–	1.7:1	39"	2
	Base Loaded Trunk Mount Antenna	vertical	–	1.7:1	39"	5
	Mobile CB– AM Antenna	vertical	–	1.8:1	46"	4
	CB Bumper Mount Antenna	vertical	–	1.6:1	102"	3
	Mobile CB– AM Antenna Converter	vertical	–	1.8:1	Reg. ant. plus 7 1/4"	2
	Center Loaded CB Gutter Clamp Ant.	vertical	–	1:1	20"	3
Mark Products	SM-27A	vertical	–	–	6'	1.5

Mfg.	Model	Type	Gain (db)	VSWR	Length	Wgt. (lbs.)
Mark Products	HW-11S-4	vertical	–	–	4'	.5
	HW-11S-6	vertical	–	–	6'	1
	HW-11-4	vertical	–	–	4'	1
	HW-11-6	vertical	–	–	6'	1.25
Master Mobile	CB-35A	vertical	unity	1.5:1	63"	–
	CB-52	vertical	unity	1.5:1	42"	–
	CB-55	vertical	unity	1.5:1	44 1/2"	–
	CB-129A	vertical	unity	1.5:1	45"	–
Mosley Electronics Inc.	Lancer 23	vertical	–	1.5:1	8' 7"	1.75
	Devant 2	vertical	–	1.5:1	45"	2
	DP-27	vertical	–	1.5:1	43 3/4"	1
	DA-27	vertical	–	1.5:1	17"	1
	SUC-1	vertical	–	1.5:1	36"	1
	PER-1	vertical	–	1.5:1	36"	1
	CBL-23	vertical	–	1.5:1	100"	3
New-Tronics Corp.	CB-111	vertical	–	–	83"	2.75
	CB-211	vertical	–	–	83"	2.75
	RM-11	mobile res.	–	–	29"	.5
	MO-1	mast	–	–	54"	2.25

MO-2	mast	–	–	54"	2.5
FGB-27-L	vertical	–	–	–	1
FG-27	vertical	–	–	–	.33
CB-27-L	vertical	–	–	26 1/2"	1.5
				60 1/2"	
CB-2	vertical	–	–	26 1/2"	.75
				60 1/2"	
TC	vertical	–	–	–	.75
TCS-27	vertical	–	–	–	.75
TCS-27-L	vertical	–	–	–	1
TCS-27-M	vertical	–	–	–	1.5
	AM/FM/CB				
TCS-2	vertical	–	–	–	.5
RTS-27-L	roof mount	–	–	30"	1
RTG-27-L	gutter clamp	–	–	25"	1
TGA-27-L	trunk groove	–	–	–	1.25
TGF-27-L	trunk groove	–	–	–	1.25
RTB-27-L	base loaded	–	–	45 3/4"	3
TMA-27	trans. ant.	–	–	–	–
CB-102-A	vertical	–	–	108"	4.5
BM-1-SPL	bumper mount	–	–	108"	5
SW-102-R	vertical	–	–	102"	1.5
SW-108-R	vertical	–	–	108"	1.5
10-3	vertical	–	1.5:1	8'	1
173	vertical	–	1.5:1	4'	.5

Shakespear
(C/P Corp.)

Mfg.	Model	Type	Gain (db)	VSWR	Length	Wgt. (lbs.)
Shakespear (C/P Corp.)	181	vertical	–	1.5:1	9'	1
	345	vertical	–	1.5:1	36"	1
	345-1	vertical	–	1.5:1	36"	1
Stinger Antennas	MM-500A	magnetic mount	4.1	1.1:25	32"	–
	Terry I	vertical	5.2	1.1:25	32"	–
	Terry II	mtch. network	–	–	–	–
	Circamount	–	6.4	1.1:0	10"	–
Thunder Industries Inc.	7010	vertical	–	1.1:1	35"	.75
	7020	vertical	–	1.1:1	36"	1
	7030	vertical	–	1.1:1	35"	.75
Thunderstik Products	48	vertical	–	–	48"	–
	100	vertical	–	–	96"	–
	T60	vertical	–	–	–	–
Versa-Tronics	CB-11	vertical	–	–	38"	–
	CB-11-1/2	vertical	–	–	20"	–
	CB-11W	vertical	–	–	37"	–
	CB-11-1/2W	vertical	–	–	19"	–

	CB-11K	vertical	—	—	38"	—
	CB-11-1/2K	vertical	—	—	20"	—
Vistaco, Inc.	Coiltenna	vertical	—	—	3"	4 oz.
	Electenna	vertical	—	—	18"	6 oz.
Waters	370-CB	vertical	4	1.1:1	108"	5
Webster	MA-100	vertical	unity	1.1:1	46"	1.5
	MA-110	vertical	unity	1.1:1	45"	1.75
	MA-120	vertical	unity	1.1:1	18"	1.5
	MA-130	vertical	unity	1.1:1	18"	1.5
	MA-300	vertical	unity	1.1:1	45"	1.5
	MA-310	vertical	unity	1.1:1	46"	1.5
	MA-500	vertical	unity	1.1:1	48"	2
	MA-510	vertical	unity	1.1:1	50"	2
	MA-530	vertical	unity	1.2:1	93"	4
	MA-540	vertical	unity	1.2:1	77"	4

MARINE ANTENNAS

Antenna Specialists	ASM-4	vertical	—	1.5:1	96"	4

Mfg.	Model	Type	Gain (db)	VSWR	Length	Wgt. (lbs.)
Antenna Specialists	ASM-23	vertical	-	1.5:1	97"	3
Apelco	M-29	vertical	-	-	54"	2
GAM	Mariner	vertical	1.5	1.1:1	19'11"	-
Mark Products	HW-11-6M	vertical	-	-	6'	4
	SMP-27-AM	vertical	-	-	6'	4
Master Mobile	CBM-16	vertical	1.5	1.5:1	79"	-
Mosley Electronics Inc.	CC-27-A	vertical	-	1.5:1	96 3/8"	2
Shakespeare (C/PCorp.)	72-1	vertical	-	1.5:1	12'	7.5
	176-1	vertical	-	1.5:1	18'6"	9
Stinger Antennas	ST-27	vertical	4.1	1.1:25	30"	-
Webster	M-29	vertical	1.4	1.5:1	60"	2

Chapter 6

Coax Cable & Connectors

The purpose of a transmission line is to deliver RF power, generated by the transmitter, to the antenna. The transmission line can be thought of as a radio-frequency hose in which RF signals flow, instead of water. If we didn't have transmission lines it would be necessary to mount the antenna directly upon the transceiver. CBers use a transmission line called "coax," short for coaxial cable. It consists of a center wire or conductor with a metallic braid woven over an insulating material which separates the braid from the center wire. The braid makes the cable more flexible and easy to handle. And it can take a lot of punishment. Excessive heat which melts the insulation, however, can cause the cable to short.

Most people who need coax cable to connect between their transceiver and antenna merely ask for a catalog number, an "RG" designation and length requirement and then wait at the dealer's counter for the order to be filled. Unfortunately, however, this procedure has produced disastrous effects upon the system performance of thousands of 27-MHz stations currently in operation. Many installations—even those with the most costly Class D CB gear—are actually producing less than one watt of output at the antenna.

Coaxial cable is not a foolproof commodity which can be bought "blind." A feedline made by one manufacturer can exhibit completely different operational characteristics from that made by another—even though both cables carry the same RG designation. Even worse, the term "RG cable" has led to considerable confusion. RG actually designates cable meeting <u>latest</u> revision specs of MIL-C-17. Older versions of JAN-spec and MIL-C-17 RC cable do not. Unless the manu-

Typical 100 ft length of coaxial cable.

facturer states this fact, it cannot be assumed that the latest spec is being represented. Additionally, some manufacturers have blurred this distinction with terms such as "RG-type." Your entire CB system could fail as a direct result of buying the wrong coax for your given application. Slow cable degradation, a prime cause of gradually-deteriorating signals, is extremely hard to detect.

USING MILITARY SPECIFICATIONS

In some instances, it might prove valuable to review military specifications. While many will be irrelevant, some reveal key parameters applicable to CB needs. For example, consider the percentage of braid coverage in a typical 27-MHz transmission system. If the cable is to prevent signal leakage that might interfere with other services (such as television), the percentage of braid cover should be quite high—at least 90% of the dielectric must be completely shielded. Yet, many cables presently being employed in this kind of RF work exhibit only 65% coverage. Add to this other problems that frequently develop, and the overall evaluation process you perform can be more clearly appreciated.

IMPORTANCE OF THE DIELECTRIC

What's important about a cable dielectric? Simply stated, the dielectric quality of any coaxial line determines both long- and short-term attenuation as well as overall power-handling capabilities. If a thick-wall coaxial line with silver-plated copper conductor has a dielectric that appears amber or gray when placed upon a sheet of white paper, it is probably com-

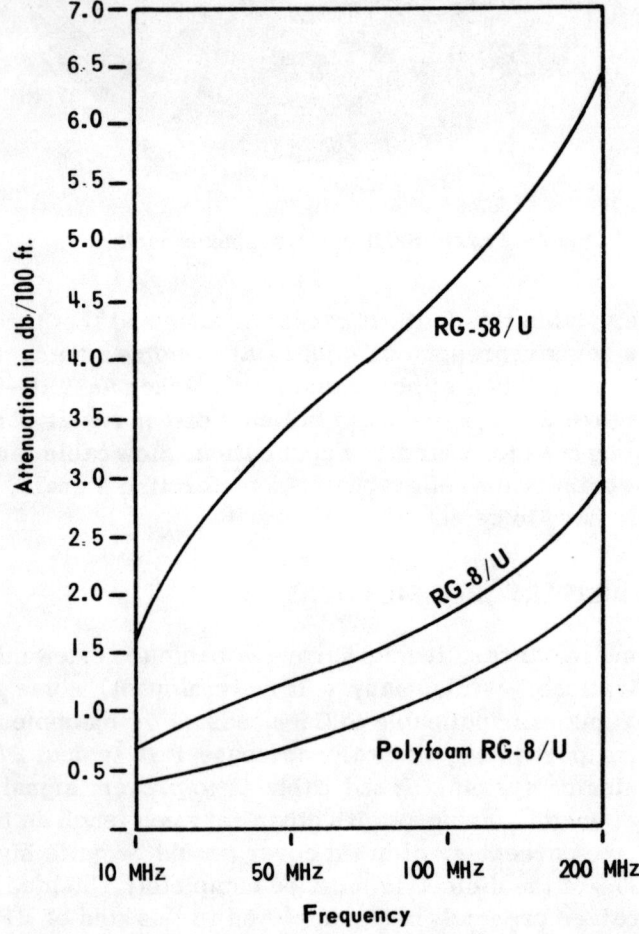

This graph shows what you stand to lose as frequency increases. Notice particularly the difference between RG-58/U and polyfoam RG-8/l.

Military RG-Number	AMPHENOL Number	Armor O.D.	Jacket O.D.	Jacket Type	Shields		Dielectric O.D.	Dielectric Type	Center Conductor	V.P. %	Capacity PFD/FT	Max. Operating Volts RMS	Nominal Impedance in Ohms	For RF Connector Series
					Inner	Outer								
8	621-111	—	.405	I	—	C	.285	FP	7/19 C	80	24.5	1500	50	UHF
11	621-100	—	.405	IIIa	—	C	.285	FP	14C	80	16.5	3000	75	UHF
11(Triax)	621-102	—	.460	IIIa	C	C	.285	FP	14C	80	16.5	3000	75	Triax
59 Type	621-715	—	.195	IIIa	—	C	.107	FP	22 CW	80	17	500	72	BNC,TNC, UHF
59	621-186	—	.242	P	—	C	.146	FP	20 CW	80	17.3	1000	75	BNC,TNC, UHF

* Signifies maximum O.D.   S — Silvered Copper   C — Copper   T — Tinned Copper   CW — Copperweld   N — Nichrome
SCW — Silver-coated Copperweld   P — Polyethylene   SSP — Semi-Solid Polyethylene   TF — Teflon   SST — Semi-Solid Teflon
TT — Teflon Tape   FP — Foamed Polyethylene

Sampling of characteristics of popular foamed polyethylene dielectric coaxial cables.

posed of inferior or scrap polyethylene. Inspect a sample of the cable you are replacing (or currently using). Demonstrating the color check can be extremely helpful. Bear in mind, however, that wall thickness, which can vary from one impedance to another (such as between 50-, 75-, and 95-ohm types), determines opacity; opacity, in turn, determines color hue. Also, conductor color can affect overall hue. Foam cables, too, can be inspected visually. Bubble size should be tight on the conductor and round in shape throughout. If a micrometer's handy, check the extrusion of whatever cable he's now using. Does the cable manufacturer monitor the o.d. and capacitance of the polyfoam core?

Asking yourself a few of the following questions can shed much light upon both the operational efficiency of your CB system and the requirements that should be met with the new cable buy. Does the line become brittle or fluid during periods of temperature extremes? Or, is it presently fluid or brittle? What about dimensional stability? Have gradual changes been noticed?

CHECK LINE CAPACITY

Standard coax lines are actually extremely long capacitors each exhibiting a pronounced effect on the tuned output circuit at each end (transmitter, antenna, etc.). To cope with this problem, coax cables are rated in terms of dielectric constants. As the constant figure approaches 1.00, the more nearly the capacity (and subsequent attenuation) approaches the low figure of open-wire lines. Knowledge of this figure allows you to analyze frequency-handling capabilities of the coax in question.

Cellular polyethylene types ("foam cable") for example, are rated at a dielectric constant of 1.5, compared to 2.26 for conventional solid polyethylene. A look at solid dielectric RG-8/U cable in terms of capacity shows an operational capacity of 29.5 pfd per ft. This compares with 24.5 pfd for foam lines of equal size. If you want to buy a 100-ft. length, the difference between one dielectric and another can be 500 pfd—enough to severely degrade matching of the most well-designed 27-MHz transmission system.

RG/U CABLE TYPE	NOMINAL ATTENUATION DB/100 FT, MHz				FOR RF CONNECTOR SERIES
	100	1000	5000	10,000	
8/U	1.9	8.0	27.0	100.0	BNC, C, HN, LT, N, Triax, UHF, Plug-In, Splice, Term.
8A/U	1.9	8.0	27.0	100.0	HN, N.
58/U	4.6	17.5	60.0	100.0	BNC, BN, C, HN, MB, MHV, N, SM, TNC, Triax, UHF, Plug-In, Splice, Term.
58A/U	4.9	24.0	83.0	100.0	BNC, MB, MHV.
59/U	3.4	12.0	42.0	100.0	BNC, BN, C, MB, MC, MHV, N, SM, TNC, Triax, UHF, Plug-In.
59A/U	3.4	12.0	42.0	100.0	Same as above.

Sampling of popular polyethylene dielectric cables, not foam type.

RF ATTENUATION

All standard coax cables have black polyvinyl chloride jackets, but there are actually two sub-categories of jacket material that should be evaluated. If you don't consider these now, you may be plagued with high attenuation in a few months.

Type I, the first category represented, is found only in older versions of JAN and MIL cables and can prove troublesome in certain applications. Depending upon age and environmental temperatures, it is possible that the polyvinyl chloride's "plasticizer" will begin to migrate out of the jacket and into the cable. Result of wrong application choice? Electrical characteristics will be drastically changed, to say nothing of attenuation. In nearly all instances, then (since temperature extremes are likely to be encountered in any uncontrolled atmosphere), you should be using Type IIa polyvinyl chloride jacketing material. Incidentally, the cost for this extra protection seldom exceeds $.02 per foot.

Both RG-8/U and RG-58/U cable should be examined at this point in terms of attenuation pattern. In the chart below, both solid-polyethylene and polyethylene-foam types are compared. The db rating is per 100 ft. lengths:

Freq.	RG-8/U Solid	RG-8/U Foam	RG-58/U Solid
10 MHz	0.55 db	0.32 db	1.25 db
27 MHz	0.96 db	0.92 db	2.40 db
50 MHz	1.33 db	0.77 db	3.13 db

These figures assume no cable degradation due to heat or general aging—under ideal circumstances such as with new Type I or Type IIa jacket lines. With a 5-watt signal and RG-58/U, more than two db of RF power input is lost at 27 MHz. With RG-8/U foam, however, less than one db (0.92 to be exact) is sacrificed. Operationally, this means you would have to generate considerably more than the maximum allowable power of a CB transmitter to achieve the same results appreciated with foam-dielectric RG-8/U.

CONDUCTORS

You should realize that in order to meet many of the standards and ratings discussed thus far, other elements of coax construction come into play. Many of these, while seemingly obscure, may account for difficulty you are experiencing now with a previous brand of cable.

Again, if a sample can be obtained of existing in-use cable, much can be determined. Visually, for example, check whether the conductor is off center in the dielectric. (If there is more than 10% error—the maximum allowed under MIL-C-17—serious problems can be expected.)

Are there as many strands in the center conductor as specified in the latest MIL spec requirement? Though this might seem unimportant, it can be crucial in 27-MHz CB work.

In most good-quality standard coax lines, braid should fit tightly. If it does not, this can indicate a strong possibility of change in electrical characteristics. Braid tightness, however, can vary; RG-8A/U, for example, has an extremely loose braid. Again, it is wise to check cable specifications.

Be certain to inquire as to flexibility requirements. Maximum flexibility is achieved with strand-type center conductors, although attenuation losses can be cut appreciably with solid conductor carriers. The answer to this question involves considerations we've been discussion.

CABLE SELECTION

In addition to the above knowledge CBers should consider the following:

Impedance: What is the actual impedance percentage of spec variation? Jacket Composition: Is it of non-contaminating material, according to MIL specifications? What are changes (in attenuation) at 27 MHz after aging tests? Jacket Tightness: Does it fit tight enough to show braid marks clearly, or is it loose, possibly indicating poor extrusion and instability? Connectors: Has the right coaxial connector been selected for use with the standard RG cable chosen? Is it designed specifically for this purpose, or does it require an adapter?

Have you seen the manufacturer's certification that the cable meets given RG specs? Does this company have a federal stock number assigned by the government to QPL suppliers? Or are all "certification sheets" merely a restatement of catalog specs? The answers to the above questions can be critically

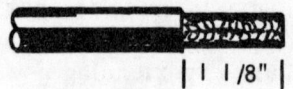

Cut end of cable even. Remove vinyl jacket 1 1/8 in.—don't nick braid.

Bare 3/4 in. of center conductor—don't nick conductor. Trim braided shield 1-1/16 in. and trim. Slide coupling ring on cable.

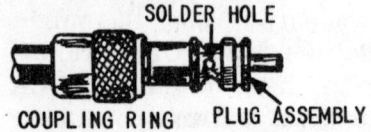

Screw the plug assembly on cable. Solder assembly to braid through solder holes. Solder conductor to contact sleeve.

Screw coupling ring on assembly.

These sketches show how to connect cables to an Amphenol 83-1SP coax plug.

important to the CBer who wants optimum performance from his communications installation. Final thought: A signal improvement of only three db translates into a more than one-and-one-half times increase in your output power. At the other end, it could make the difference between being heard or not!

CONNECTORS

When attaching a PL-259 (Amphenol 83-ISP) connector to heavy coax such as RG-8A/U, the jacket must be cut back 1 1/8", and the inner conductor bared for 3/4". The shield should be cut back to 1/16" from the edge of the inner dielectric. The coupling ring should be removed from the connector and slid onto the coax in the proper direction. Next, the outer shield and center conductor must be tinned with solder. Screw the cable into the plug body until the coax is seated. Some braid will be exposed through the hole in the plug body and should be soldered to the plug body. The center conductor is soldered to contact the hollow pin on the plug. Screwing the coupling ring back into position completes the job.

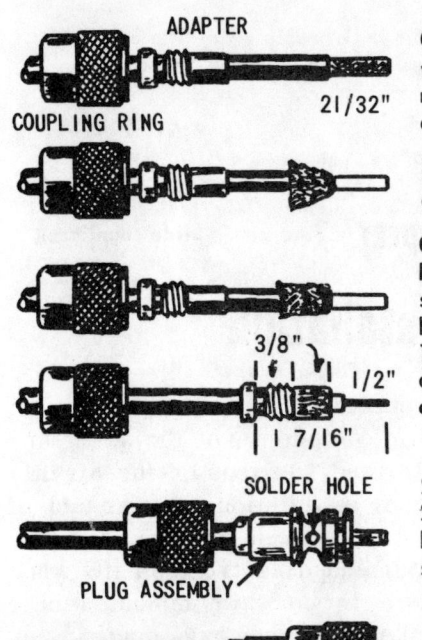

Cut end of cable even. Remove vinyl jacket 21/32 in. —don't nick braid. Slide coupling ring and adapter on cable.

Fan braid slightly and fold back over cable.

Compress braid around cable. Position adapter to dimension shown. Press braid down over body of adapter and trim. Bare 1/2 in. of center conductor— don't nick conductor. Pre-tin exposed center conductor.

Screw the plug assembly on adapter. Solder braid to shell through solder holes. Solder conductor to contact sleeve.

Screw coupling ring on back shell.

Here, special adapters (Amphenol 83-168 and 83-185) are used to attach a coax plug to a cable.

When PL-259 connectors are used with a small diameter cable such as RG-58A/U, a poor job usually results unless a standard adapter is used. These adapters are low in cost and really simplify the job tremendously. Our diagram shows such an adapter on RG-58/U coax. The braid has been exposed for a distance of about 3/4", frayed and pushed back, and the center wire bared for 1/2" and tinned. Again, as with the heavy cable, put the coupling ring on the coax and screw the adapter assembly into the plug body. A final test with an ohmmeter should be made on all connector assemblies. A great deal of trouble can be caused by only a small piece of wire braid twisted in the wrong direction.

Chapter 7

The Installation

Most problems arising in CB operation, surprisingly, are not traceable to the selection of bad equipment, or the wrong antenna for a given situation. Instead, serious on-the-air difficulties are attributable to poor installation or utter lack of periodic CB system equipment maintenance. The reason is simple: CBers are seldom experienced electronics buffs when they are just getting their feet wet in two-way communications. To save you the grief of mistakes thousands have made in recent years, this Chapter presents a few basic hints and kinks regarding equipment installation. If you follow these tips, you can be sure of reliable CB communications over greater distances with your new equipment.

MOBILE INSTALLATIONS

Mounting CB equipment requires little more than a screwdriver, hand drill, wrench, and pliers, plus a few moments of your time. But it can prove a disaster if it's approached "cold" by the rank beginner bent on getting his equipment on the air the fastest way possible. Most CB transceivers are designed to operate from 12-volt auto batteries, with the negative pole of the battery grounded. Lift the hood and check to make sure that the battery terminal marked "Neg" or "-" has a braided metal strap connecting it to the auto body. As long as the car or truck is of at least 1960 American vintage or later, this probably will be the case. Foreign cars have varied electrical systems, so always check carefully before proceeding. Assuming the car has a 12-volt negative-ground system, the next step is that of selecting a place under the dash for mounting the transceiver. The unit should not be po-

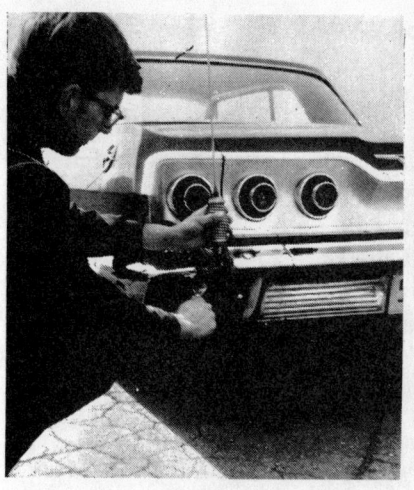 

The antenna installation should be made first. Although the rebuilt '63 Chevy Super Sport shown here used a combination antenna (see other photos which follow), the installer demonstrates how a standard bumper mount should be installed. A hole should be drilled and grommeted to permit coax to enter the trunk. In picture above, installer puts final touches on the half-wave whip mount by tightening the two adjustable spacers. Completed bumper-mount installation (r). Notice that coax passes through the drilled hole, not fed through normal trunk opening, where damage to the cable can result, eventually short-circuiting the feedline.

sitioned directly in the path of the heater air stream, because temperature extremes can affect frequency stability, and excessive heat often damages components. Crystal microphones are particularly susceptible to damage by high temperatures.

An important factor to consider in equipment installation is safety of operation. The unit must be mounted in such a way that it does not interfere with proper operation of the vehicle. If it is located beneath the dash but too close to the steering column, the transceiver could interfere with brake-pedal travel or cause the driver difficulty when applying the brakes. Just as dangerous is a transceiver location too far to the right, forcing the driver to lean over to operate it. For the same reasons it is best to use a microphone with a push-to-talk button.

The underside of the dashboard is not always level: often it is cluttered with heater controls, cigarette lighters, ash trays, accessories, etc. In these cases, there are several mounting alternatives. You can simply choose a different location, move the obstruction, or use mounting brackets that extend well below the interfering projection. Some CBers use the hump in the center of the front floorboard for mounting equipment. In any case, the microphone hanger can be fastened to the dash within easy reach of the operator. Self-tapping screws are used to fasten the unit to the floorboard.

Next step is to trace the mounting holes in the bracket onto the underneath of the dash. (It may require some gymnastics to see these tracings and drill the holes.) You'll want to lie down a little to the right of the driver's seat, so that the small of your back is where the seat of your pants usually rests.

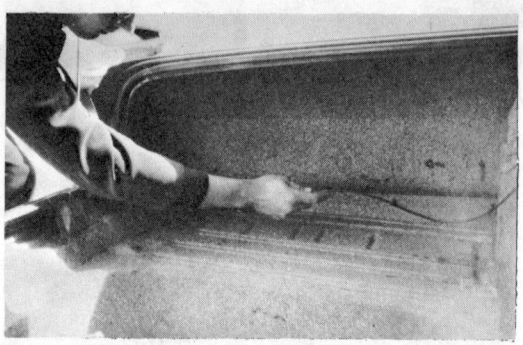

Inside trunk, cablig (RG-58A/U) is routed through channel to opening behind car seat (far right, not visible). Cables should always be routed in corners and crevices to prevent entanglements.

Cabling routed under carpet. No drilling is necessary here. Notice that slack has been provided; otherwise a break could occur when rear seat is replaced.

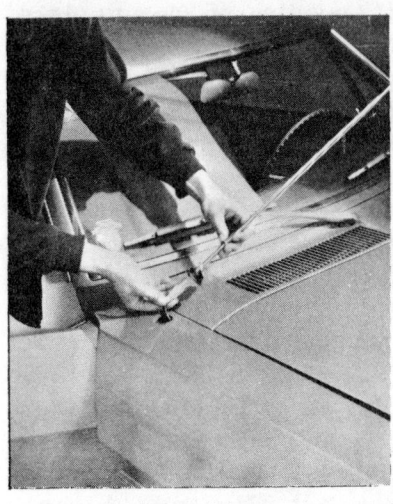

On the car an Antenna Specialists M-103 combination AM/CB center-loaded whip is used. In photo above left, original radio antenna has been removed. New coax line, furnished complete with required fittings, is routed through opening and connected to base of AM/CB antenna. This can be accomplished prior to locking the mount in place. To install, only a wrench is required. The mount locks automatically, precluding the need for underside adjustment or two-hand installation.

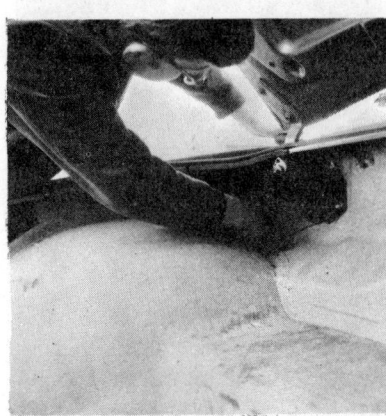
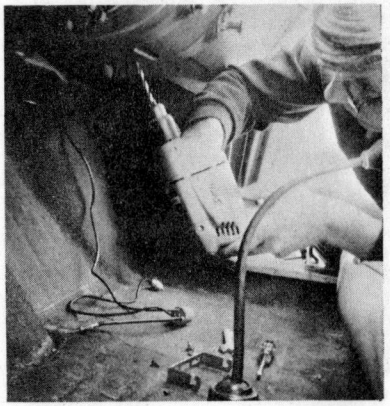

Inside trunk/hood of the car the installer takes up slack in feedline. Cable end with PL-259 fitting is now in the driver's compartment under dash. Next step is to drill at least two support holes for the CB transceiver bracket. Self-tapping screws should be used, with drill size only large enough to permit initial threading. A center punch should be used to position the drill bit; otherwise drill may veer off the target, scratching the finish.

Although somewhat ungainly, this position provides a clear view of what is happening and leaves both hands free for tools. With a center punch or a large nail make a clear indentation in the middle of each mounting-hole mark under the dash. This gives the electric drill bit something to seat into, and practically eliminates the chance of slippage.

Next, put on a pair of clear goggles; the type used for skin-diving will do if you don't have one of the clear types used for protecting the eyes when spray-painting. While drilling, tiny shavings of metal will spiral away from the bit and fall on your upturned face. Only goggles which seal the eyes from particles entering from the sides are safe. You'll want to drill the punch-marked holes to the correct size with a 1/4" or larger chuck drill equipped with the appropriate bit. Be careful not to let the drill cut through the metal too fast. Use only enough pressure to keep the bit against the dash. No more.

When the holes have been drilled clean and burrs removed with a file or tapered reamer, bolt your mounting bracket in

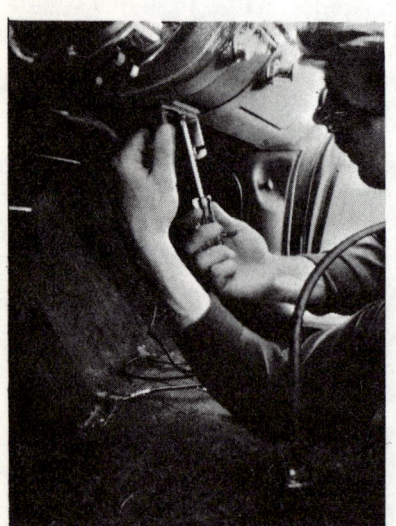 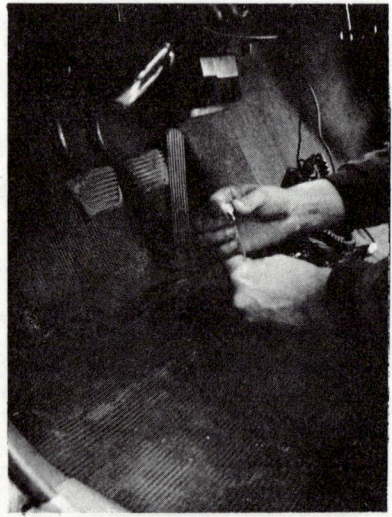

installation of the bracket. In some cases, holes will be found under astray, precluding necessity for drillihg. Always position the transceiver for operational convenience. You won't want the microphone cord to become entangled with the steering wheel..which occurs more often than you might believe. Next, a push-on clip can be affixed to the red lead (12V DC line) from the CB transceiver. Standard push-on clips designed to mate with auto clips used at termination points under the dashboard, are available from most automotive supply houses.

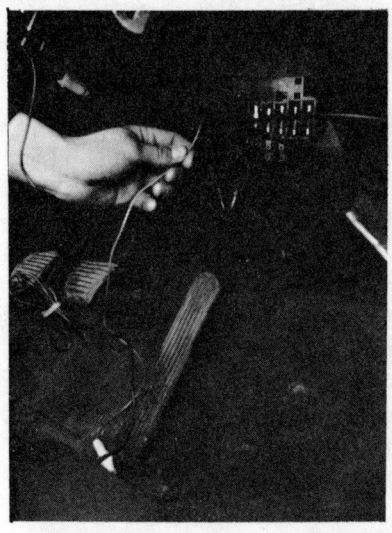 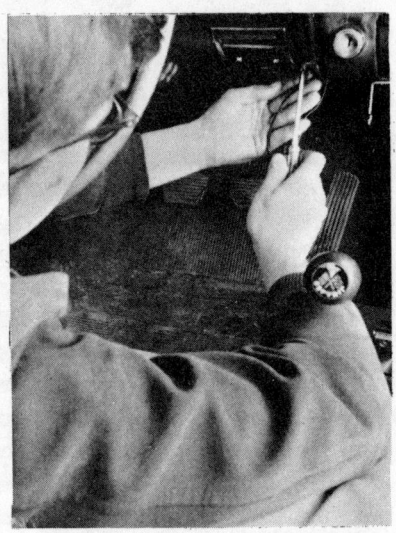

Next, the red lead is pushed onto unfused "accessory" terminal which is electrically connected to the battery. This termination point is generally found under the dash against the firewall and can be used when the transceiver power lead is fused. Otherwise, you can terminate directly to the accessory terminal at the ignition switch. Next, the ground (black) lead from the transceiver is connected to the car by means of a lug inserted under a loosened Phillips-head screw (r). Always make this connection, regardless of transceiver mount. The ground normally achieved through a tap-mount is not sufficient in most cases. Disastrous results have been produced under conditions of severe vibration—particularly with the newer, all solid-state units.

place. Be certain to use lockwashers under each bolt head and nut to prevent loosening from vibration. Install the transceiver into the bracket, taking care to make sure that it is correctly placed.

Now you'll want to bring power to the CB transceiver. With older (tube-type) units, it is recommended that you wire the power lead in series with the ignition switch. Reason: If the transceiver is accidentally left on, the tube filaments will not run down the car battery while the engine is turned off. Connecting the transceiver power lead directly to the ignition switch is not desirable since the switch wiring usually has high resistance and resulting power losses cause a corresponding reduction in transmitted signals than if the transceiver were connected directly to the battery. Transistorized units have such low battery drain that it makes little sense to worry about accidentally leaving the transceiver on when the engine isn't running.

Nearly all CB transceivers come equipped with a female power plug for connecting the transceiver to the battery. One lead on the plug is usually black, the other red. Ground the un-fused lead to the body by wrapping it tightly around one of the bracket mounting bolts, and screw the nut down firmly to keep it in place. Trace the dash wiring to the firewall to find the hole through which you can run the power lead from the battery to the transceiver. Then, splice a No. 16 or larger rubber-insulated wire to the fused lead from the power plug, long enough to reach the positive battery terminal. Don't just wrap the wires together—solder and tape the splice to prevent power losses or accidental short-circuits.

At the car battery end of the No. 16 wire, form a 1/4" diameter loop and coat it with solder for sake of rigidity. Then unscrew the bolt on the positive battery lead, remove the nut, slip the loop on the exposed bolt end and replace the nut. Make certain the bolt is tightened firmly, then coat the battery terminal, battery lead, and power lead with petroleum jelly.

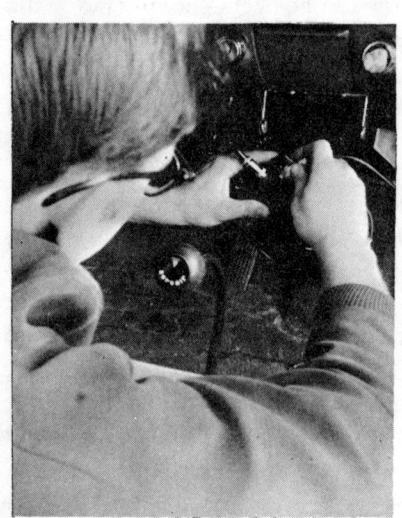 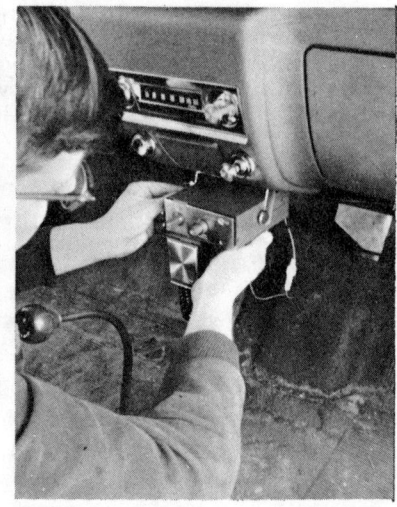

Next, the PL-259 connector on feedline is screwed onto the transceiver's SO-239 antenna fitting. Notice that all connections are made prior to actual unit installation. Now the unit can be attached to the mounting bracket. Always use the transceiver mount recommended or provided by the manufacturer.

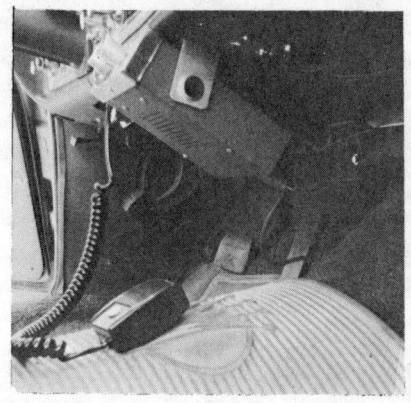 

At this point, power and gound leads are pushed back out of view. Notice that coax enters through opening in carpeting. Combination mike/speaker can be clipped on the side of the transceiver as shown or pushed into position just below the unit. Where you install your mike clip (r) should also be evaluated from safety angle; never should mike or cord interfere with normal driving operations.

Driver pulls out, putting call into his base station five miles distant. Notice the clean, overall appearance of installation. Total time involved? Only 54 minutes.

The transceiver is now electrically connected and needs only an antenna.

MOBILE ANTENNAS

Mobile antennas for a CB radio system should be chosen with care if maximum efficiency is to be expected. Appearance is a major concern to some CBers. For example, the user may want inconspicuous antennas for company cars, etc. Salesmen as another example, may be limited by leasing company contracts which forbid cutting a hole in the auto body for mounting an antenna. In any case it's worth thinking about, since CB antennas can tend to be large and obtrusive.

The irregular shape of an automobile dictates only one location where the antenna should be mounted: square in the middle of the roof. Antenna length and aesthetics may rule out this location, but it is practically the only one where radiation patterns are unaffected by the car's steel body. The antenna against which most all others are checked is the quar-

If at all practical CB, antennas should be mounted in the center of the roof.

Here is an example of a difficult installation problem where air conditioner precluded positioning of the CB unit closer to the driver. In such cases, the unit should be operated only when vehicle is at standstill. Otherwise, cord and controls would force distraction of driver's attention from roadway.

ter-wave whip. (In some cases, however, it is the dipole.) While the whip is usable at most frequencies above 20 MHz, it is interesting to notice that at 27 MHz it is 108 inches long. This limits it to either bumper or fender mounting.

By using a loading stub in the feedline, the quarter-wave whip can be shortened to about 54 inches. Since it is shorter than a full-size whip, it can be mounted on top of the rear fender to improve the radiation pattern. The shortening stub is cut to the correct length so that a wattmeter in series with the antenna and transmitter reads minimum reflected power. Where a degree of gain is desired and it is physically possible, an excellent omnidirectional antenna for CB frequencies is the base-loaded rooftop whip. An inductor and a capacitor in the sealed cylinder at the antenna base combine to match the shortened antenna to the transmission line.

In truck cab installation, the transceiver mounting bracket is intalled behind the seat and positioned to place transceiver operating control within easy operating reach. In this installation, the microphone bracket is installed above the rear window; this keeps the coiled mike cable clear of the steering wheel.

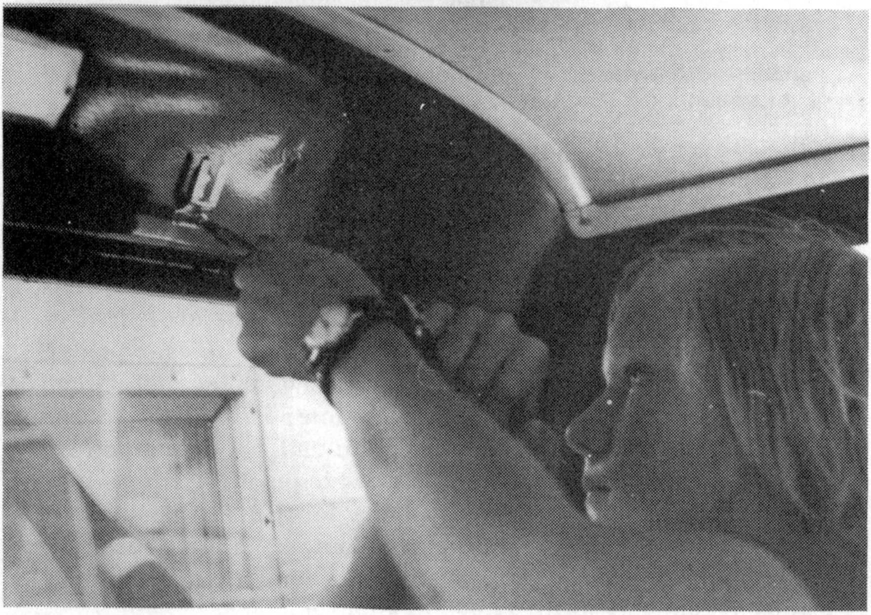

At right, a "truck mirror" type of antenna mount is installed on the top mirror support arm. Below, the antenna cable shield is soldered to a terminal lug and secured under one of the antenna mounting bracket studs. Be sure you have a good ground and metal-to-metal contact.

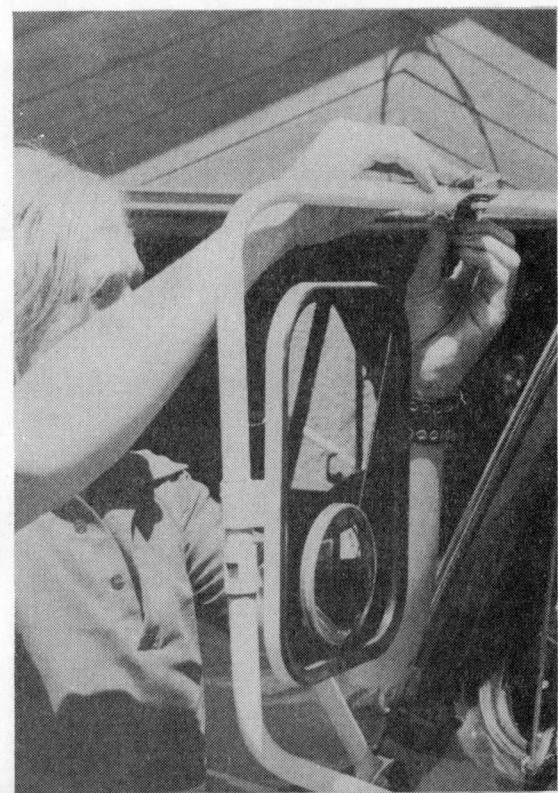

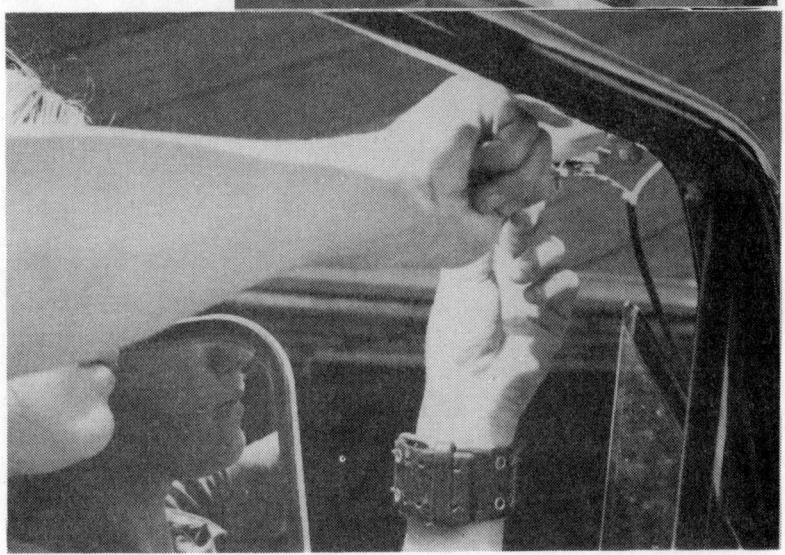

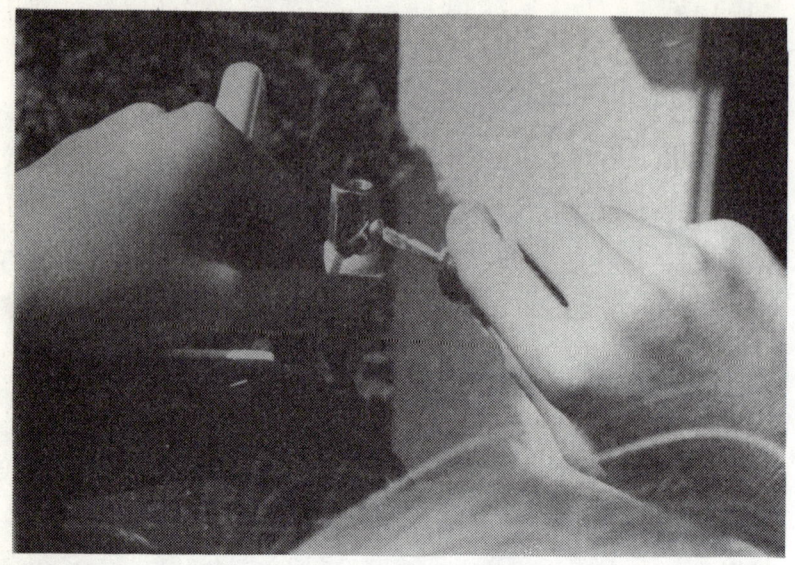

A

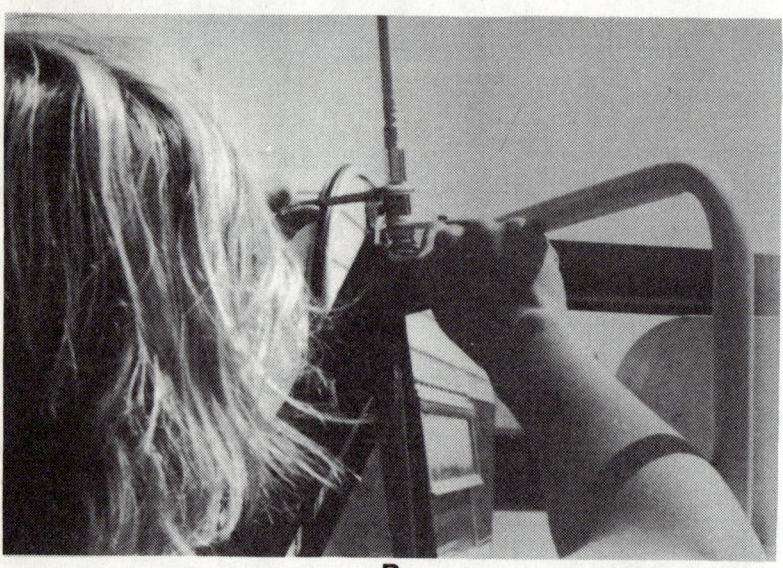

B

Antenna cable center lead is connected to the insulated antenna mounting stud. Antenna is attached to the mounting stud and then securely tightened using two wrenches to prevent the stud from turning on the bracket.

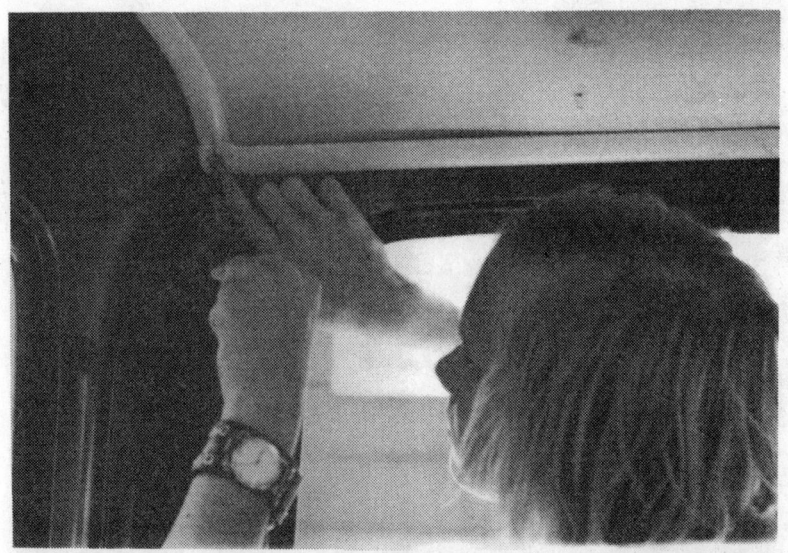

The coaxial cable from the transceiver is routed up to the roof headliner and secured with small plastic cable clamps using the screws in the molding. The coax can also be routed under the molding above the door to the front of the cab. The cable goes out through a grommet and hole in the roof to the antenna on the mirror or simply out the front of the door. If you don't want to drill a cable hole in the cab, the rubber insulation around the cab door frame is generally thick enough to prevent pinching the coax. However, be sure to check this first. Remember to leave enough slack in the coax to allow for opening the truck door.

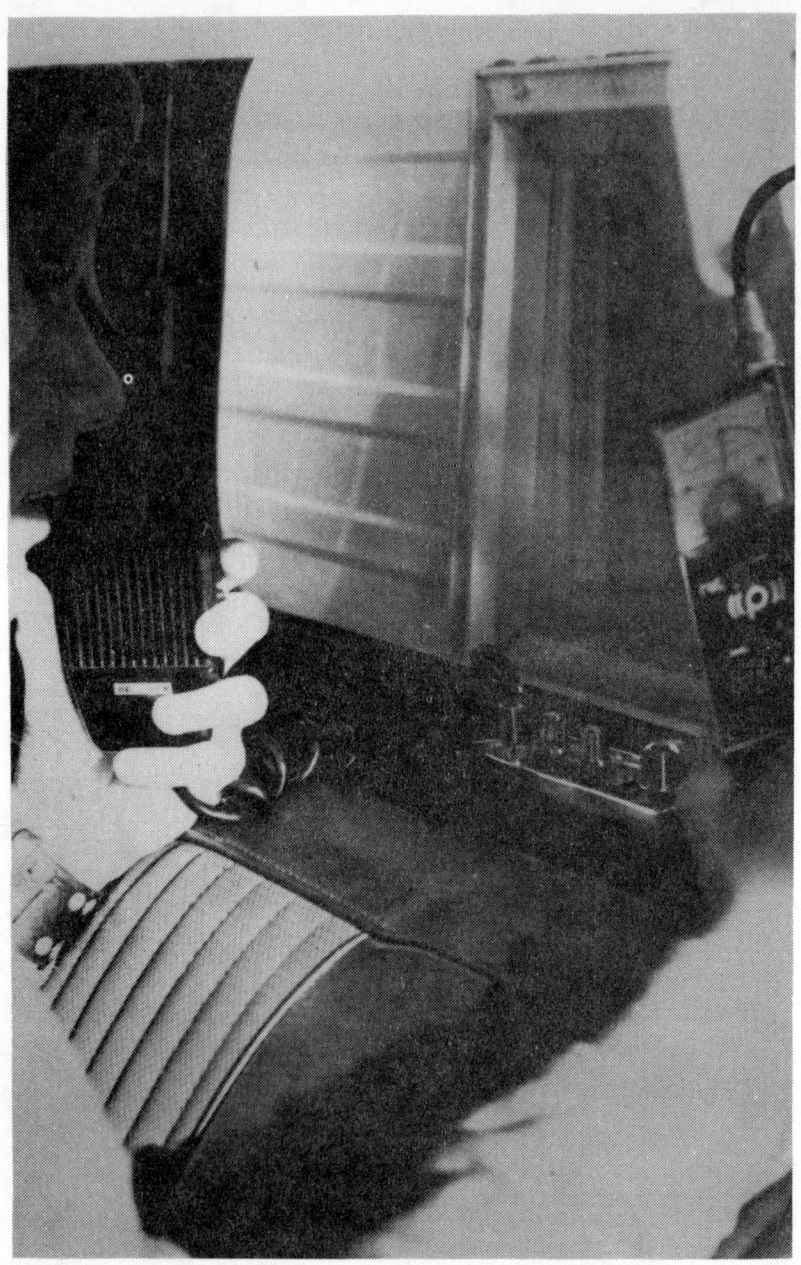

Once the installation is completed, it's a good idea to use an SWR meter to check the forward and reflected power. A low reflected reading (1.2 or less) and a high forward reading indicate that the antenna system is operating properly. The finished installation can best be checked for signal readability, modulation, and noise by actual on-the-air operation.

At 27 MHz the base-loaded whip is long, but it has a bandwidth of about 5.0 MHz. Because it is roof mounted, and because it delivers a signal 50% stronger at any given distance than the whip, it is an excellent general-purpose mobile antenna. A variation of the base-loaded antenna is the helically wound fiber-glass whip. Basically a large radiating coil, the helical antenna can be made in varying lengths for a single resonant frequency. It offers less length than a whip at equivalent frequencies, and its efficiency is about the same. By increasing the length of a stub-loaded whip to 5/8 wavelength, omnidirectional gain of about 2.5 db will be secured. Like the unity-gain model, the 5/8 wavelength whip can be either cowl or fender mounted.

A 5/8 wavelength gain antenna is available without the stub loading. This antenna combines the characteristics of base-loaded rooftop and stub-loaded models. A loading coil is formed into the base of the spring-tempered stainless steel whip, resulting in a gain of about 2.5 db over a simple quarter-wave ship. Additionally, adapted versions of many of the above antennas are now available as car-radio-antenna replacements. While not ideal, they should be considered. (See Chapter 5 on antennas.)

NOISE SUPPRESSION

Following are some tips for eliminating noise caused by various components of the vehicles's ignition system. None of these should impair normal operation of the car.

Auto Gauges: Place a .01-mfd disc capacitor between each gauge terminal and the nearest grounding point on the car's frame. Also do this at the ignition switch.

Distributor: Install a distributor suppressor from the center terminal of the distributor. These are available from most auto and CB dealers.

Spark Plugs: Install new Auto Lite or Champion resistor spark plugs, or purchase add-on spark suppressors for each of the existing plugs. These are available from most auto supply stores and CB dealers.

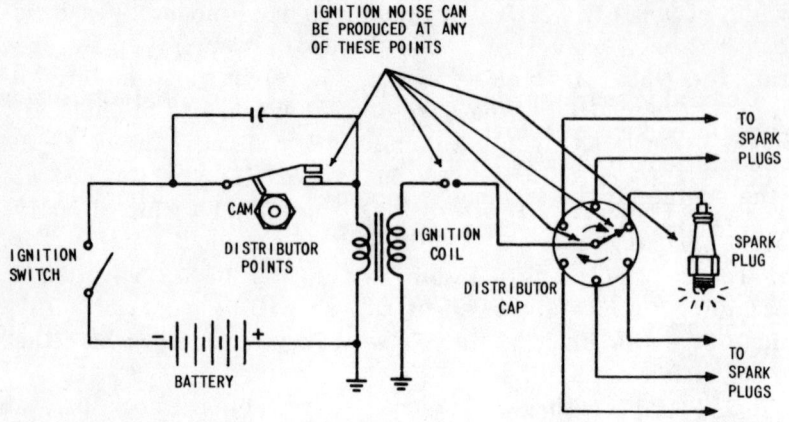

This schematic shows the electrical relationship of various components of vehicle's ignition system.

Voltage Regulator: Install a 0.2-mfd coaxial capacitor on the battery terminal marked "B" of the regulator.

Generator: Install a 0.2-mfd coaxial bypass capacitor between the generator's armature terminal and any nearby grounding point on the engine block or firewall. Good tunable generator noise traps are available in electronics mail order catalogs.

Coil: Install a .001-mfd disc ceramic capacitor with short leads from the battery terminal of the coil to the case. Next, get some bonding braid and run it from the firewall to the engine block, touching the coil case along the way. Make certain that there is a good connection at each point of contact and keep the braid as short as possible.

Motors: Install 0.5-mfd capacitors between the motor cable and ground at the defroster, electric windshield wiper, or heater motor, and any other accessory motors.

Alternator: Install an alternator noise filter, available for well under $4.00 from most mail-order electronic supply houses as well as some automotive supply houses.

Tires: Insert anti-static powder by means of an air pressure hose at a service station. Place static collector springs inside the front wheel grease retainer cups.

Coaxial transmission line for most CB mobile installations should be kept as short as possible. At the same time, coax should be routed away from the gauges, switches, relays, and the engine. If the antenna is mounted on the back bumper, the cable should be run through the inside of the car. From the transceiver run the cable along the edge of the floor, beneath the carpet and through a small hole in order to feed into the trunk area. Another hole may have to be drilled so that the line can pass from the trunk area and through a rubber grommet to the antenna. Always check for an existing opening before drilling. Always fit a drilled hole with the rubber grommet referred to earlier to prevent the cable from being damaged.

It is possible to check a mobile antenna with an ohmmeter; at least a simple continuity check can be made if it is suspected that the antenna or lead-in are shorted or defective. Connect one ohmmeter lead to the antenna tip and the other to the lead-in that plugs into the transceiver. Long leads for the ohmmeter may be used effectively, depending upon the installation. Set the ohmmeter to its lowest scale. Then shake the antenna and lead-in. Resistance should be about two to six ohms. If resistance is considerably higher, or if it varies when either the antenna or the lead-in from the antenna are moved, there is a poor connection between the antenna and lead-in. Disconnect the lead-in from the antenna and check both items individually to locate the fault, although the lead-in connector is almost invariably the most likely culprit.

Connect one ohmmeter lead to the antenna tip and the other to a ground. Shake the antenna and lead-in, with the ohmmeter set to its highest range. The ohmmeter should indicate an open circuit. If there is a constant high-resistance reading, there is a high-resistance short between the antenna and the auto body. If an intermittent short is indicated, a visual check will usually pinpoint the cause.

Finally, connect one ohmmeter lead to the outer conductor at the transceiver end, and the opposite lead to ground. Set the ohmmeter to its lowest scale and shake the lead-in. The resistance should be zero or near zero. If it is higher, the outer conductor is not making proper contact with the ground.

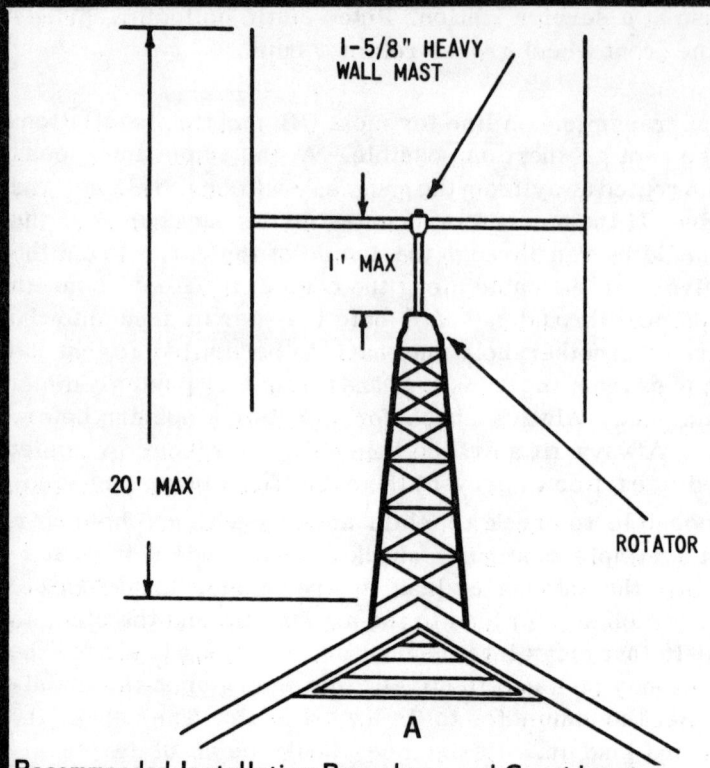

Recommended Installation Procedures and Considerations
— Courtesy Hi-Gain Antenna Co.

When installing this antenna there are several items to be taken into consideration. The close proximity of metallic objects can deleteriously affect the performance of any antenna. For best results it is recommended that the antenna be mounted at least 20 feet away from power lines, TV antennas, or any existing metallic objects. The recommended installation method, using a self-supporting tower, is shown at A. If the tower requires guy wires, they must be broken up with strain insulators every 3 feet. Although not recommended, the antenna may be mounted on the roof as shown at B above. However, optimum results may not be obtained due to metallic objects such as rain gutters, house wiring, etc. If this method of installation is used, the mast material must be 1 ⅝-inch HEAVY WALLED material. Mount the antenna a maximum of 1 foot above the rotor and guy the mast directly below the rotor. Use three guy wires equally spaced and break them up

This same procedure should be followed on base station antennas.

BASE STATION INSTALLATION

The antenna is the key to your base station installation. Once

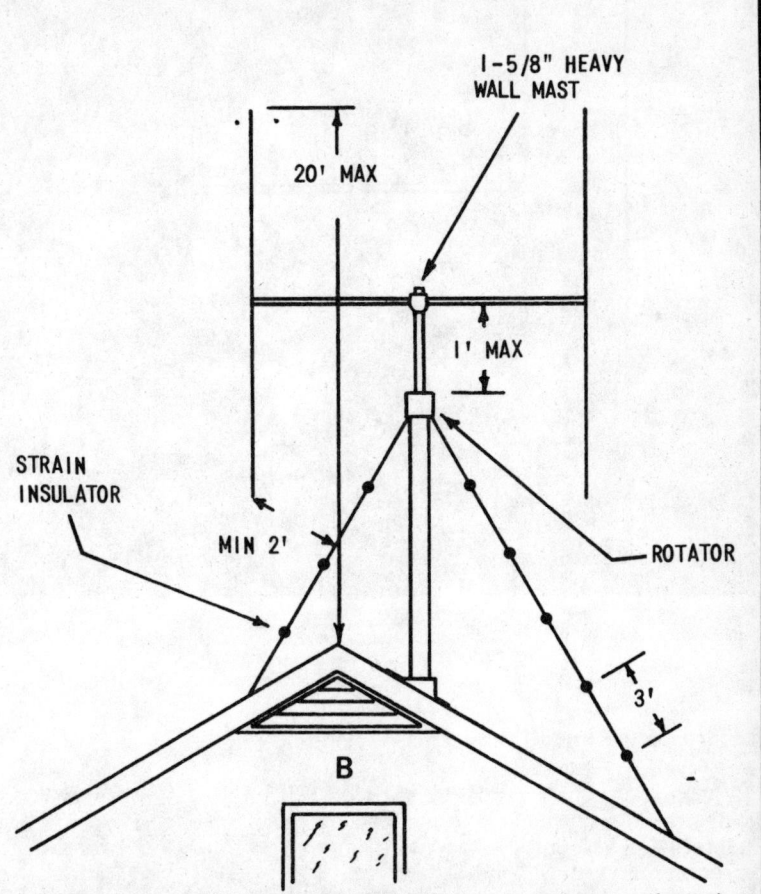

every 3 feet with strain insulators. Install the guy wires so there is a minimum of 2 feet of clearance btween the guys and the lower tip of the antenna element. This minimum of 2 feet clearance and the use of strain insulators applies to ALL installation methods using guy wires. For proper lightning protection and to insure noise-free performance, the supporting structure must be properly grounded through the use of a ½-inch by 8-foot ground rod driven into the ground as close as possible to the base of the supporting structure. Connect the ground rod to the supporting structure using no. 10 or larger copper or aluminum wire.

the antenna is in place, there is little involved in getting a base CB rig on the air. Key to best performance, of course, is in selecting the best location for the antenna. Interestingly, while the highest possible spot is normally preferable, sometimes it is also the least practical.

Here are several unusual CB installations: (A) a transceiver installed on a battery-run vehicle used for plant protection at Service Combustion Corp.; (B) motorcycle-mounted CB; (C) CB set used by KIRO DJs to communicate with volunteers traveling Seattle's freeways during rush

hours; (D) glider—believe it or not—containing CB gear (insert shows the transceiver mounted vertically inside cockpit).

Questions you might ask yourself at this point: Will your structure secure the antenna adequately in high winds? Is the base station antenna surrounded by other high objects, such as trees, which might wipe against the antenna in high wind? Is it a safe location, from an installation and maintenance standpoint? CB antennas should be guyed securely to withstand the rigors of wind, driving rain, etc. Instructions appropriate to installation which are included with the antenna selected should be followed exactly. Many must be "weatherproofed" after the coaxial connection has been made; in such cases, the manufacturer usually includes a rubber insulatory material in the form of a liquid or spray.

Another paramount factor in achieving optimum performance is cable length. Your base antenna should be mounted at a point that will afford maximum practical elevation yet still be a reasonable distance from the CB equipment itself in terms of cable power loss. Unfortunately, CB radio antennas are ungainly when it comes to actual installation. A two-man team is the minimum required and some planning is mandatory.

Experienced CBers are aware of the FCC prohibition of the use of towers to elevate Citizens Band antennas excessively. However, there is some misinterpretation of the Part 95 rule that permits the use of higher transmitting towers that have been licensed under other services. Contrary to widespread belief, this rule does not permit the use of towers used to support home television antennas. True, the tower does not come under the classification of a specially built structure for supporting a CB antenna, since its original intention is improved TV viewing. But it is not a transmitting tower. Licensed towers permitted to support a Citizens Band antenna are those which have been registered by the FCC for elevating a non-CB radio transmitting antenna. This provision is especially important to commercial and municipal users of CB who already have some other type of two-way radio antenna. (See FCC Rules & Regulations, Part 95.37.)

Subparagraph (c) of Part 95.37 states that the antenna mounted on another licensed transmitting-antenna structure must not exceed 60 feet height above ground no matter what the height of the other station's actual supporting structure.

If, for example, a small community has a police FM system operating on 39 MHz, it can install a CB antenna on the same tower for increased range. It is this provision which makes

CB equipment valuable not only as the sole method of communications for a business, but also as a supplementary radio system. By adding CB to its business/public safety licensed system, a two-way radio user can effectively add 23 channels of communications to his already licensed VHF system. And the additional height offered by the tower will result in solid coverage over a much greater area than normal communications radios.

Time is your most important factor. How much time will it take to locate a proper mounting point? How much time will it take to assemble the antenna? How much drilling must be done? Are guy wire points easily accessible, or will they have to be provided, thus eating away still more time? How many men can be recruited for the job? How much time will this take? How much time must be spent fastening connectors and soldering on new ones? What about holding points for the cables? If a rotator is used, how much more time will its addition to the installation add? Why the emphasis on time? Simply because it can prove the most critical governing factor; if pressed against it, results can be disastrous, both from performance and safety standpoints.

A word about grounds. All CB installations demand a good groung connection, which can be provided by driving a standard copper ground rod into the earth and running a wire (not less than 12 gauge) between it and the antenna support pole. Another lead should run to the chassis of the equipment itself. Never omit this step in your installation.

Always inspect the antenna and lead-in at regular intervals for obvious damage from corrosion at the coaxial terminals, abrasions, etc. More important, your antenna should maintain a minimum SWR over the range of frequencies on which it will be operated. Periodic SWR checks are in order. Many CB stations use an SWR meter connected in the antenna lead-in at all times, providing a constant check of antenna operation.

Chapter 8

Optimizing Your Station

The prime purpose of your base station is to communicate with, and to direct the other units in your system. The best way to make this an efficient operation is to have the station equipment within easy reach, convenient, and comfortable. Accessories such as tone signaling or selective-calling equipment should be placed close to the transmitter where they can be activated if necessary.

Most CB stations utilize a push-to-talk microphone, but you might consider rewiring it to a foot switch to allow freedom of both hands. It is often a simple matter to parallel the switching contacts to a foot switch for hands-off operation. Take a good look at your mike, too. Some mikes are better than others and some have more output. Since the audio is used to modulate your transmitter, it is good practice to be sure that your voice signals are accurately reproduced and strong enough to provide normal modulation. Many transmitters have built-in speech compressors to insure the proper audio level at all times. If your transceiver doesn't have speech compression, there are accessory models available that are easy to install.

Another practical and popular addition to a base station is a second antenna for a particular application. This can be done by routing your antenna lead-ins through a coaxial switch to the transceiver. Normally, these coaxial switches have a provision for connecting the transceiver to one of several antennas. You might well consider having one output on the coaxial switch connected to a matched dummy load for easy, accurate transmitter testing.

All of your AC-operated base-station equipment should be fused and connected through the master switch to the AC line.

This not only affords safer operation, but, if necessary, you can quickly close down all line-operated equipment at one time. Be sure all AC equipment is well grounded. One way to do this is to use 3-wire grounded plugs and circuits in your base station. Another way is to run a heavy wire from your transceiver to a nearby cold water pipe (preferably a copper pipe). After scraping the surface to provide a clean contact, clamp the wire in place. Then connect all other AC-operated equipment to this common ground point on the transceiver. If you don't have a water pipe or 3-wire ground plugs handy at your location, you can purchase a relatively inexpensive copper ground rod. Drive it into the earth where it will provide for the shortest length of wire to connect it to your equipment.

Other items which will add to your base station operating ease are a telephone, phone directory, pencils and holder, scratch pads, maps of the local area, emergency phone numbers, and of course, spare fuses and tubes.

MOBILE SYSTEM IMPROVEMENTS AND ACCESSORIES

Mobile operation is slightly more critical than that of a base station because you have some variables and restrictions to contend with that you don't have in a base system. Some of these are vibration, temperature variations, changing supply voltages, moisture, dust, corrosion, noise, limited space, and restricted operating procedures.

Because of these factors, you want to take all the precautions available to get the most out of your rig. The best time to apply these considerations is during installation, of course, but the following suggestions may reveal the cause of an inefficient or improperly operating set. Make sure the proper wire size is used for connections to the DC source. Secure mounting is vital to prevent vibration, and the coaxial cable must be carefully routed from the transceiver to the antenna. Also, nothing is more important than proper installation of the antenna. Once your mobile rig is installed and checked out, you might want to know how to make this basic system just a little more effective.

POWER MONITORS

Since mobile operation is more critical and your antenna height relatively low, you want to be sure all of the available

This is a typical low-pass filter (cover removed). Properly designed, these filters can be placed in the feedline between the CB transceiver and antenna and adjusted for minimum attenuation of transmit signal. Their big attraction: they can also be tuned to nearly eliminate TVI (television interference).

power is radiated. This means constant monitoring of the power output and VSWR. There are accessory units for this which are connected in series with the coaxial cable and draw no power. They are ideally suited to mobile use and can be left in the system to warn you if your antenna system or transmitter go sour. This is important because a mobile antenna can become damaged, coaxial cable can be cut, or the transmitter can become weak due to low voltage. You can use this power monitor/VSWR indicator to tune your antenna for maximum output, greater range, and insured continued optimum performance.

SPEECH COMPRESSOR

Modulation capabilities vary somewhat with different transmitters, but in AM transmission, the percentage of modulation has a great deal to do with the amount of power your transmitter puts out. As indicated in Chapter 3, the power in AM transmission is divided in three parts—the carrier

Equipment shown here is usually sufficient for most on-the-air monitoring. Above the transceiver is a direct-reading wattmeter; at right, an inline CB transceiver checker capable of reading SWR, among other things.

Here is a close look at a piece of selective-calling equipment. The plug at left connects to the back of the CB set. This system used vacuum tubes. Today's modern types use solid-state devices.

and the two sidebands. The sidebands contain the modulation or amplified voice signals, so get all the power you legally can into these sidebands. The type of microphone and a variation in voice can also affect the amount of modulation. A man with a low, powerful voice talking into a mike with good low-frequency response can easily modulate a transmitter to its full capacity or over! A person whose voice is higher and softer will probably not modulate the rig to full power.

To get around this, many companies design their transmitter audio section with built-in speech compressors. A speech compressor raises a softer input signal and lowers a strong input signal to provide a fixed amount of modulation at all times. This means that the speech compressor is set to give 100% modulation for constant, optimum output regardless of how loud or soft the operator talks.

SELECTIVE CALLING

This is one of the most effective CB operating accessory

units designed to increase range and positive primary contact. It is especially useful in areas of crowded channels. Selective calling is a system where audio tones are used to modulate a transmitter, and because such tones can be closely selected, only compatible receivers will hear them.

It works like this. The tone-signaling device consists of an audio oscillator using a tone reed which vibrates at a specific frequency when it is turned on. The tone oscillator is wired to the transceiver for its operating voltage. A switch on the tone-signaling device connects the audio tones to the transmitter modulator during the time it is keyed. (It actually takes the place of your microphone with the tones replacing your voice.) When the signals reach a similarly equipped transceiver, the received tones can be identified. In some selective-calling systems the received tones are used to actuate an alarm circuit to let the station operator know he is being called. Once the system has performed its job of primary contact, the station operators can call each other and assume normal communications.

PERIODIC CHECKS: 12V DC SUPPLY

Supply voltage variations in mobile operation can cause some drastic variations in performance. This effect is more pronounced in tube units which receive their filament voltage from the vehicle battery. A tube with a weak filament, for example, can lose almost all of its output with a 2- or 3-volt drop in

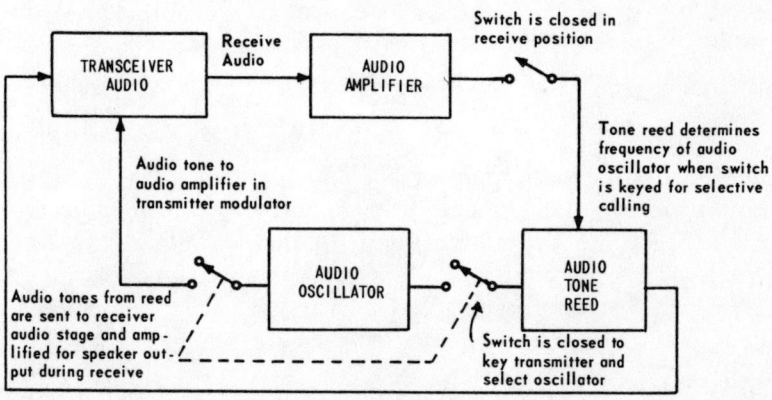

Block diagram of a selective-calling unit.

filament voltage. Transistorized equipment is usually voltage-stabilized with zener diodes to prevent overload. The performance of solid-state equipment is normally not adversely affected by minor drops in voltage since no tube filaments are involved.

One way to insure adequate supply voltage and maximum operating efficiency is to be sure your battery and voltage regulator are in proper condition. Periodically clean the battery connections with a mixture of water and baking soda.

MOISTURE, DUST, AND CORROSION

Getting rid of these little gremlins comes under the heading of preventive maintenance, and to make sure your mobile station is kept in good shape a periodic cleaning is in order. Every few months remove the transceiver from the cabinet and clean out the dust. A little contact spray cleaner on the channel selector switch and in the volume and squelch controls will prevent noisy, intermittent operation. If your rig has a relay for switching the antenna and audio circuits during transmit, it would help to clean them with a relay contact file. Use a file made for this purpose or you may destroy the contact plating and spacing adjustment. If your CB antenna is mounted where it is exposed to road dirt and grime, periodically remove and clean it. Corrosion and grime can affect the radiation properties of the antenna, the VSWR — and your operating range.

Chapter 9

Troubleshooting Your Rig

Your CB license gives you privileges as an operator of CB equipment, but it does not make you a technician. As a matter of record, many CB operators do become interested enough in electronics to pursue it and become licensed technicians. However, as a non-technician, there are a few service repairs you can legally perform on your equipment. There are adjustments in the transmitter that you are not allowed to fool with except as a properly licensed technician or under the direct supervision of one. A properly licensed technician is one who holds a valid 1st or 2nd class commercial radio operator's license. In this case, a 1st or 2nd class commercial radio telephone.

The following troubleshooting charts indicate a symptom and some of the possible causes you can check. Most of the CB equipment today is transistorized and a few general service hints will aid in troubleshooting. The following charts are designed to help you quickly locate the general causes of malfunction. But, if it is necessary to dig into the circuit, restrict your servicing to non-critical areas unless you really know what the effect will be on circuit operation. Never attempt to adjust or replace parts in the transmitter that could change the frequency tolerances or the modulating capabilities.

We don't recommend you start tearing into a transceiver unless you know what you're doing, have the test equipment, a schematic, and know how to use them. It's one thing to change tubes and fuses, but quite another to search out and replace components in solid-state circuits. If you have doubts about repairing a certain malfunction, by all means take the unit to a licensed, qualified repair station.

Technician with First Class Radiotelephone license (at least Second Class required by FCC) makes transmitter checks.

Underside of a typical, tube-type CB transceiver. This unit has 10 crystal sockets (for 5-channel transmit/receive). Notice that one receive and one transmit crystal are in place. Exercise extreme caution when working on sets with power applied.

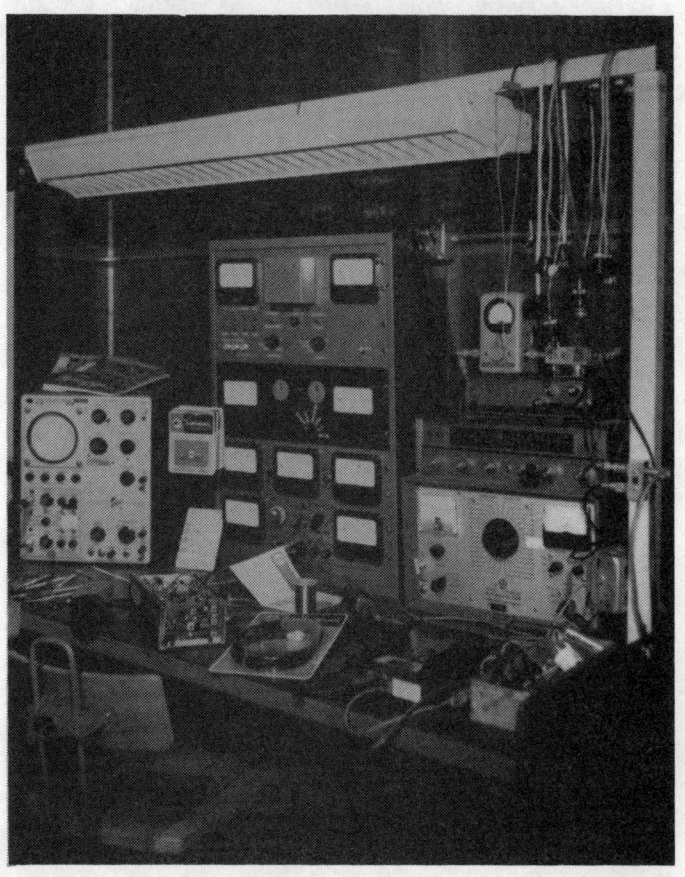

Why take your CB transceiver to the shop for servicing? A look at the elaborate equipment required for the job tells the story. This installation is at Broadview, Illinois.

If you are inclined and equipped to dig out your own circuit problems, checking transistor voltages is a good way to isolate a defective component if you have a schematic handy. If a transistor emitter voltage is high, or there is very little difference between the emitter and base voltages, you can suspect a shorted bypass capacitor or an open emitter resistor. A low emitter voltage can indicate an open transformer, defective transistor or a defective resistor. A resistor can increase in value or become overheated and crack (carbon resistors seldom short out).

If the collector voltage is low or about the same as the emit-

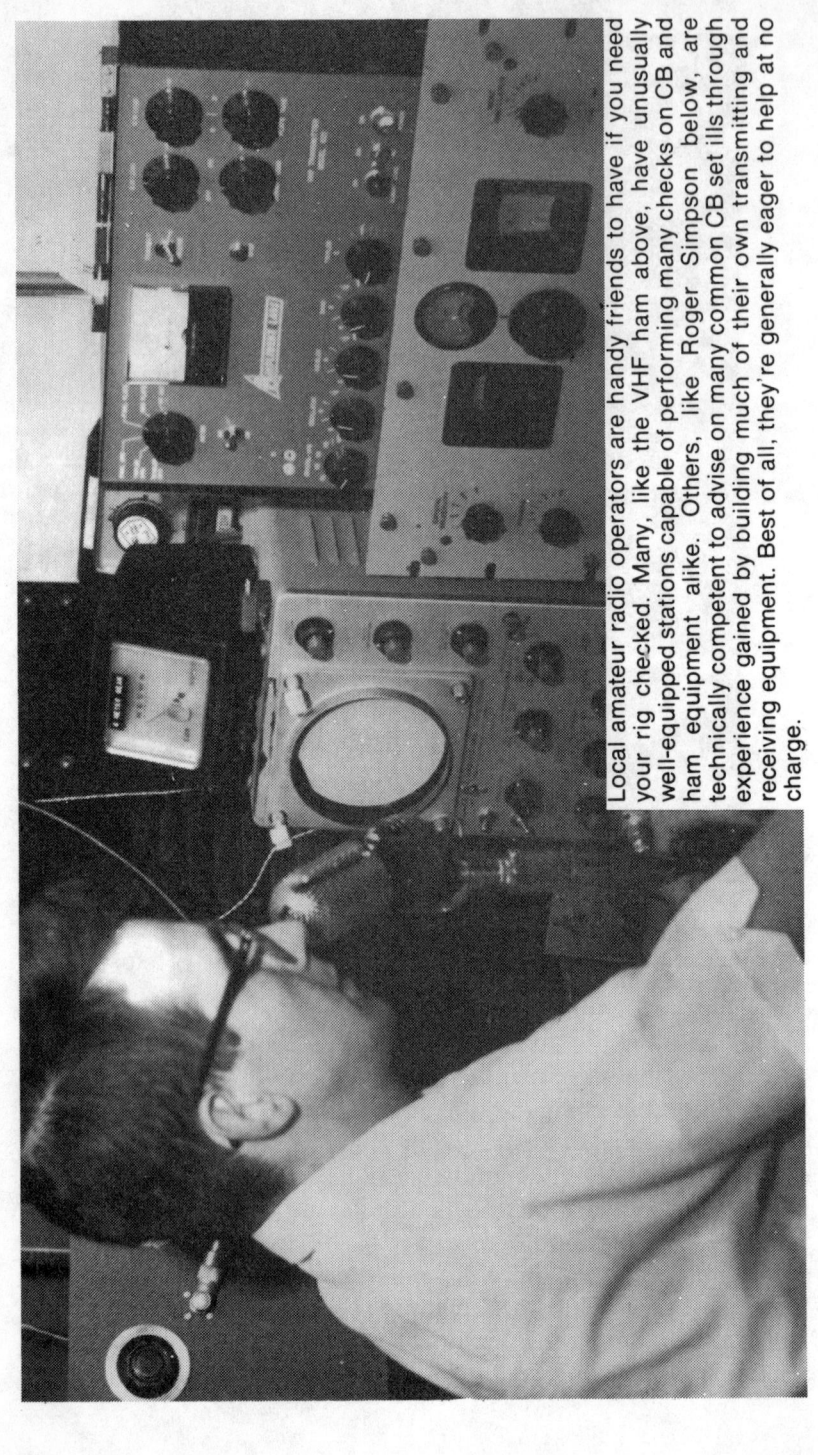

Local amateur radio operators are handy friends to have if you need your rig checked. Many, like the VHF ham above, have unusually well-equipped stations capable of performing many checks on CB and ham equipment alike. Others, like Roger Simpson below, are technically competent to advise on many common CB set ills through experience gained by building much of their own transmitting and receiving equipment. Best of all, they're generally eager to help at no charge.

ter, you probably have a shorted or very leaky transistor. Zero voltage on the collector may indicate an open transformer or decoupling resistor. In a defective stage, check for proper base and collector bias. If these values check normal, check transformers, capacitors, resistors, and transistors. Also check the printed-circuit board for broken or cracked lands.

When soldering or removing diodes and transistors, use a low wattage iron, 30 or 40 watts, and a suitable heat sink. The heat sink can be an alligator clip on the lead you are heating between the body of the component and the soldered connection. This will absorb the heat and prevent damage to the component.

GENERAL TROUBLESHOOTING

Symptom	Possible Cause
Transceiver dead	Check the fuse and power cord connections at the unit and at the source of voltage. Check grounding. The cabinet and mounting bracket may be the only ground in the mobile. Check to see if tubes or dial lamps light; if so, the unit is getting power but possibly no B-plus voltage.
Fuse blows	Check to be sure the fuse is the correct rating specified by the manufacturer. Check the power source for proper polarity, especially in mobile units. If the transceiver uses a vibrator power supply, the vibrator may be sticking. Replace it. Check power connections to be sure the hot lead is not shorted to ground. Remove the unit from the cabinet and check for shorts in the capacitors connected between the power bus lines and ground. Check the power lead through the on-off switch.

RECEIVER TROUBLESHOOTING

Symptom	Possible Cause
No noise from the speaker	Check power connections and fuse as indicated for a dead transceiver. Check the on-off switch to be sure it is on. Are the volume and squelch controls properly set? Check the receiver on other channels. Try a new receiver crystal. Check speaker connections. If the unit has an external speaker plugged in, be sure that connections are correct and that the external speaker is good. Check for an open speaker. Remove the speaker, it should read continuity across the terminals. Check for dirty or sticking relay contacts, especially if one is used to mute the speaker. Check for an open audio detector diode. Check tubes or transistors in the squelch and audio circuits. If the transmitter operates normally, you may be able to eliminate problems in some of the audio stages as some of these stages also are used for modulation. Check the receiver oscillator. If the receiver and transmitter use frequency synthesized units, a defective stage in the synthesizing network would affect both receive and transmit functions. The oscillators can be heard on a communications receiver with its antenna lead placed near the oscillator. Check for an open volume control. An open volume control can be located by touching your finger on the lead to the audio stage. If you hear a loud buzz with your finger on that end of the control but not on the other tap(s), the control is either open or turned all the way off. Check the circuit board for loose solder connections and broken or cracked leads. Check transistor voltages.

Symptom	Possible Cause
Distorted audio	Check audio modulation on transmit. If the modulation seems normal, you can eliminate some of the audio stages in your unit. The speaker may be torn or dirty. Check and replace if necessary. Check AVC circuits, audio output transistors, audio coupling capacitors, and tubes or transistors in audio circuits. Unit may need alignment—see note below.
Noise, but no signals received	Check the channel selector switch. Replace the receiver crystal if operation seems normal on other channels. If the same symptoms appear on all channels, check the oscillator. Check the antenna switching relay for dirty, sticky, or misadjusted contacts.
Weak Reception	Check the volume control setting. Check the antenna and switching relay. Check the receiver crystal. Check tubes or transistor voltages. Unit may be out of alignment—see note below.
Intermittent operation	Check the volume control for worn spots. Turn it back and forth slowly to see if audio cuts off. Check the channel selector for dirty switch contacts. Check relay contacts for wear, dirt, or misadjustment. Sometimes the spring tension on relay contacts gets weak and they do not make good connections. Check power connections. A plug can have loose-fitting pins or terminals. Check the antenna lead-in connector. If the unit is tube-operated, gently tap the tubes to check for intermittent elements. Remove the chassis and check for loose or broken wires and components. If the transceiver is solid-state, check the circuit board

Symptom	Possible Cause
	for cold solder joints, cracked lands, and open connections.
Receiver hum	Check filter capacitors in the power supply. Check the volume control.

NOTE: If the receiver needs alignment, have it serviced by a properly licensed technician.

TRANSMITTER TROUBLESHOOTING

Transmitter dead, receiver normal	Check the transmitter on other channels. If it operates on other channels, replace the crystal in the dead channel. Also check the channel selector switch for open or dirty contacts. If the transmitter is dead on all channels, check the push-to-talk relay. Check tubes or transistors in the transmitter oscillator and power output stages. Check voltage in these stages.
Power output, no modulation	Check the receiver audio section. If normal, replace the mike and check for modulation. Check relay contacts. If both receiver and transmitter use the same audio stages and the receiver audio is weak, check the tubes or transistors and voltages in those stages common to both sections. Check for shorted or open modulation transformer windings.
Power output, weak modulation	Try a new microphone. Check voltages. Unit may be out of alignment. If transformers or transistors were replaced in the transmitter, it may need retuning— by a properly licensed technician.

Chapter 10

CB & Public Service: Providing Emergency Assistance

One of the greatest assets to having a CB rig under the dash of your car is the fact that it offers you instant emergency aid—help when you want it. You always know that all it takes is a push of the mike button to raise a REACT monitoring station (see Chapter 1).

Did you ever stop to think about those who come to your call for help? Did you ever think about the planning and work that went into the fact that they were there when you wanted them? Did you ever wonder if you might be able to offer this same service to others? Probably not.

True, it does take a pinch or two of planning and a bit of elbow grease; but the feeling of satisfaction you get when you help out your first stranded motorist, or donate your club's services to the solution of a problem facing the community (flood, lost child, police assistance, even handling the messages involved in a parade) can be well worth the effort. Actually, all you need to get the whole thing off the ground is a desire to do it plus a few other public-spirited CBers. Once you've accomplished this, your best bet is to contact REACT (Radio Emergency Associated Citizens Team), a well-known national organization which encourages CB public service activities.

REACT isn't a "CB club"; that is to say, it is not concerned with any social aspects of "CB'ing" such as card swapping, coffee breaks, jamborees, etc. It's a non-profit group which isn't particularly concerned with the internal politics or organization of local affiliated groups (known as "teams").

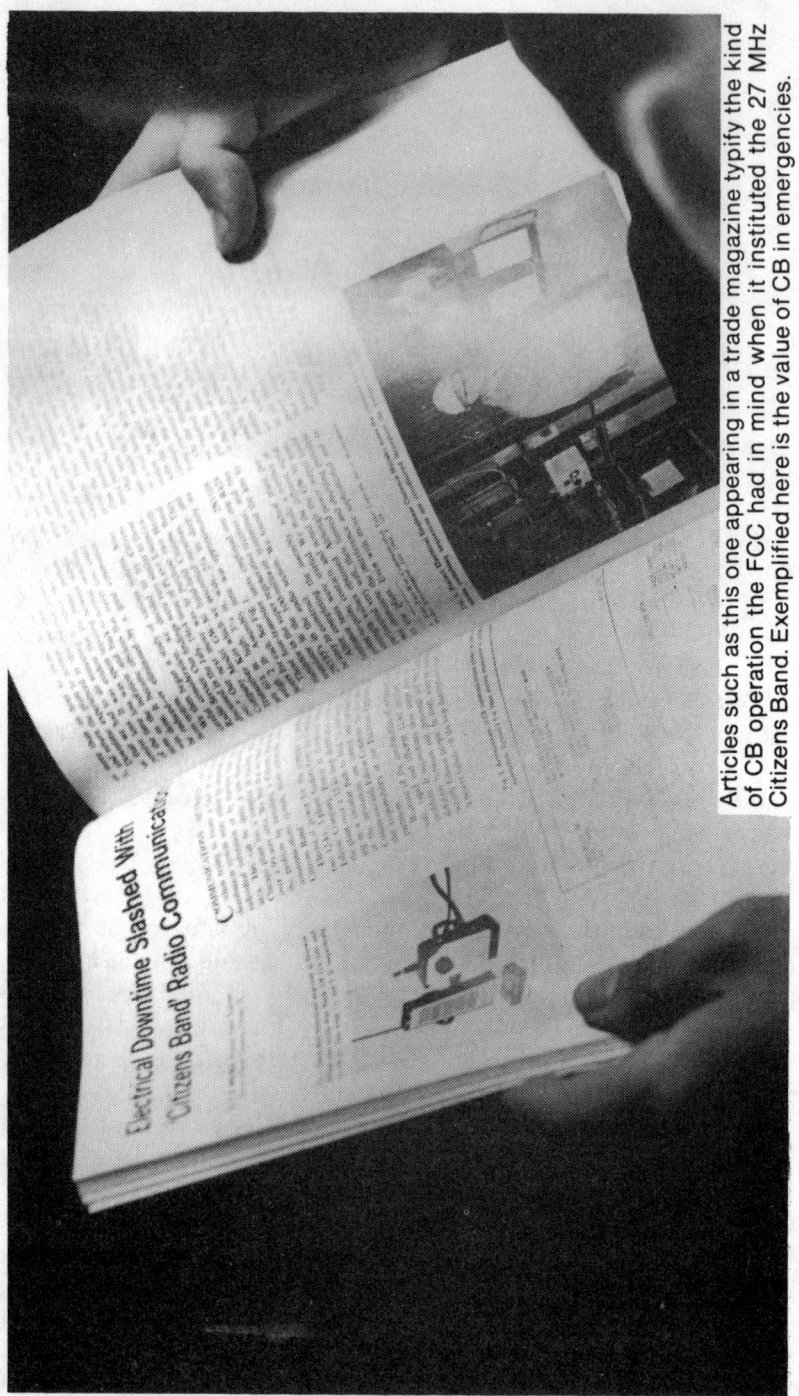

Articles such as this one appearing in a trade magazine typify the kind of CB operation the FCC had in mind when it instituted the 27 MHz Citizens Band. Exemplified here is the value of CB in emergencies.

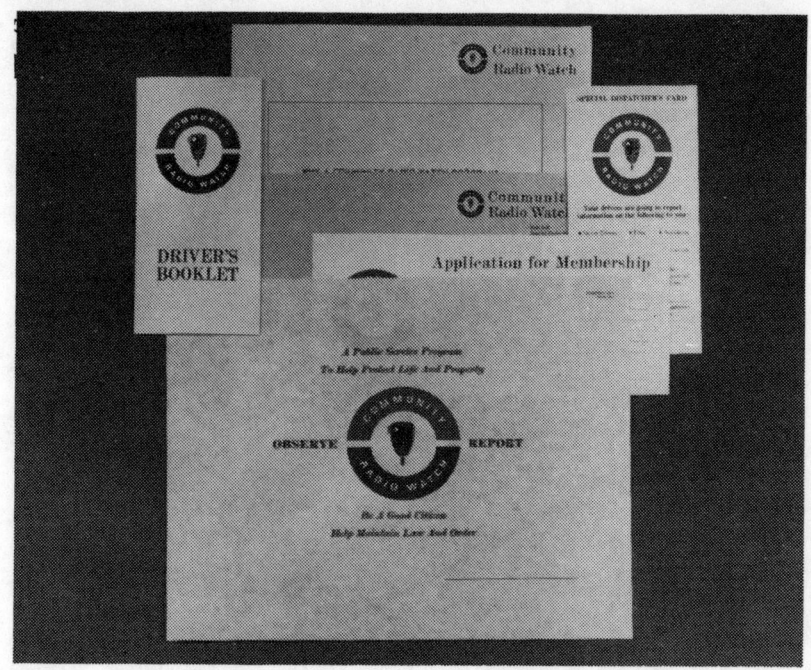

Community Radio Watch program materials.

REACT statistics based on a recent system survey:

Organization:

Number of participating teams	1100
Average number of members per team	27
Total of active membership	30,000
Total CBers with REACT training	100,000

Messages:

Percentage of calls for road assistance	52.0%
Percentage of calls on actual accidents	21.4%
Total emergency situations handled monthly	50,238
Total emergency situations yearly	602,856

After several years of REACT's non-club activities, they have racked up impressive numbers of members and 1100 teams, and the amount of reserve operators (CBers with REACT operational knowledge) is well in excess of 100,000. The function of REACT is to advise these teams of the latest techniques and theories in efficiently serving the community via 27-MHz radio, setting up emergency monitoring stations.

In rural and farm areas, emergency monitoring can be of particular significance. Even in dead of winter, a stranded motorist on a small country road five miles away can be assured of a response on Channel 9. This home has monitoring transceivers both in the house and in the barn (notice ground plane atop silo).

coordinating with police and fire authorities, proper operating procedures, etc. Our suggestion is to fire off a letter to REACT National Headquarters, 111 East Wacker Drive, Chicago, IL 60601. You'll receive full details on this group, together with application forms.

While affiliation with REACT or other such organizations has been proven to be an asset to any CB emergency monitoring group, it's not impossible to do a creditable job as an independent. Biggest problem has been the lack of information available on working it out. We hope, however, that the material which follows will help fill that void.

YOUR EMERGENCY MONITORING POST

Perhaps more than anything else, <u>listeners</u> are needed for CB emergency frequency monitoring. Though this hardly seems like much of a problem, the fact is that very few CBers are inclined to do much listening. They'd much rather talk. Yet, those who rely on their under-the-dash transceivers for emergency assistance certainly deserve serious attention.

Nearly half of the frantic "emergencies" handled by CB operators could be avoided entirely or greatly simplified if operators would just take the time to listen carefully before grabbing the mike button. Unfortunately, however, most don't. And the result is a chaos of clobbering, "in-the-mud" transmissions, and unrendered emergency service. If you're sincere, though, about lending a hand, we'd suggest that you begin with your base station. A real CB emergency-communications monitoring setup can perform a true service if it is intelligently put together and utilized to its utmost.

Channel 11 is no longer an official calling channel in the U.S. Channel 9 is the emergency channel. The calling channel was to be used only for establishing a contact; but it didn't work during the few months it was an official FCC rule. So it was dropped in late 1976. The only other time Channel 9 should be used is for communications relating to an accident, such as would happen if you heard a plea for help and responded, although short transmissions (also of a service nature) giving route information and the like are permitted. Although this general agreement has all the weight of the FCC behind it, you may hear uninformed CBers

In a Davenport, Iowa flood disaster CB came to forefront as both 5-watt and portable Class D units were pulled into action as secondary backup units to law enforcement officials. However infrequently they may be used, all hand-held transceivers should be priodically checked for battery condition. In an emergency, a fresh battery in your unit may well save someone's life.

break the rule by using the channel for personal non-essential communication, but this does by no means imply that Channel 9 is not functioning as an emergency in your area. It simply means that they either don't know about its purpose, or clearly do not appreciate its value. In any case, ignore these violations. The more who fail to respond, the sooner these people will switch to more "active" channels, leaving the frequency clear for legtimate emergency use

If you are serious about maintaining a top-notch communications post, you'll want to have the means for constantly monitoring 9 without disrupting normal station activity. While it's always good to have a standby transceiver hanging around the shack in case of minor disaster with the rig, it's rather expensive to go out and purchase a new transceiver which has no other function than to monitor 9. So why not look around for a reliable second-hand rig? Some of the earlier tube sets designed for only one or two operational channels can be purchased dirt cheap, cleaned up, and tied with a coaxial "tee" connector to the main antenna feedline. The magazines abound with inexpensive squelch circuits (if your second-hand transceiver doesn't have one) you can put together, and several manufacturers sell "add-on" adapters that will do the same job the $200 sets do in quieting the receiver when signals aren't present.

The concept of a second transceiver isn't the only possibility, of course, but it affords you the frequently needed capability of being able to "talk-in" motoring CBers to the scene of the accident on a local club frequency while maintaining contact with the calling party on Channel 9. But you can get by with a receiver alone.

Should you elect to buy a separate receiver, look for a good all-purpose communications receiver rather than a transceiver with it's transmitter out of commission. Besides the advantage of being able to tune the international shortwave broadcast bands, you can monitor the Coast Guard channels, the 2182 marine distress frequency, the special emergency channels, and a host of other important service frequencies. Many's the time an alert CBer was able to help safety agencies by bringing an emergency to their attention.

Another receiver you should consider a "must" is an inexpensive "monitor receiver" such as those being manufactured by Allied, Lafayette, Regency, etc. These are special VHF

Smokie's got ears! It's true, not only do half the autos seem to have CB these days, so does the Fuzz. This Pennsylvania state trooper's cruiser uses gutter clamp antenna for quick change from "Tijuana Taxi" to his personal 4-wheeler.

Florida's State Conservation Department maintains a network of CB radios between base station, automobile (left), and its fleet of boats. In addition CB channels are monitored in rough weather to assist vessels in distress.

configurations that make FM tuning over the police and fire services possible, plus allowing you to eavesdrop on numerous special services (such as medical, ambulance, etc.). The receivers come in two versions—one for the 30 to 50-MHz band, the other for 150 to 174-MHz. Some companies, however, sell receivers tuning both ranges. (Contrary to common belief, it is not illegal to monitor police bands, although in certain locales regulations exist prohibiting police receivers in vehicles not connected with official law-enforcement duties.) All in all, then, the ideal monitoring post would have at its disposal two CB rigs, a communications shortwave receiver, and one or more special FM monitor receivers.

With this equipment at hand, the next thing to find out is where to listen for emergency transmissions not received on CB Channel 9. Naturally, selection of the single channel used in your area is entirely up to the FCC and local authorities. If you haven't the time to check them all by trial and error, place a telephone call to the local or state police, fire, utility station, etc., and simply ask what frequency band they use and query the dispatcher as to the exact channel. If you state why you'd like to know and your affiliation with a local CB club, you're in the ballpark. Still another possibility, however, exists with local area two-way radio communications sales/service companies, which can be found in the Yellow Pages. These outfits are responsible for the installation and maintenance of the very radios you wish to monitor, and many of the technicians they employ are also hams and CBers.

A great timesaver, though, are the Police/Fire Radio Directories recently made available to interested CBers by Communications Research Bureau, P.O. Box 56, Commack, L.I., N.Y. 11725. Available at only a few dollars each, the Radio Directories list call signs, locations, frequencies of state, county, municipal police and fire department dispatches in major metropolitan areas (main city and surrounding communities). Each directory lists hundreds of such stations.

Though it can be fascinating to just eavesdrop on these stations, if you are seriously contemplating helping those in need, there are two more items that should be handy: A telephone and a list of essential phone numbers. This phone book should include numbers of monitoring stations in nearby areas which might be closer to the station in need of help, power and light company, hospital, state police, highway patrol, local police, sheriff, local REACT, tow truck, doctor, first aid squad, etc.

No one expects you to monitor for 24 hours a day. Yet when you do listen at your command post is vital to smooth operation and an "even" distribution of aid and road assistance. For this reason, you will probably want to work out some kind of schedule with other local CBers, preferably members of your own radio club. In addition, housewives can frequently be relied upon to put in an hour of monitoring during the day as they iron, fix dinner, etc. You'll be surprised how much coverage can be extended when you solicit cooperation with other stations.

YOUR MOBILE COMMUNICATIONS CENTER

If you are fortunate to live in an area boasting a prosperous CB club, you probably enjoy the benefits of a full-fledged emergency communications van. What this means is that somebody has raised enough money to purchase an old bus, milk truck, or delivery van and then equipped it with a half ton of radio gear and base station antennas. In the event of a natural disaster such as flooding, or a man-made emergency such as power failure, the emergency van can be rushed to the scene where it serves as a temporary base station—the central point for dispatching mobile units, walkie-talkies, etc., participating in the emergency service program.

Of course, not every club has such a van, and very few individual CBers are financially well-off enough to foot the bill for such an endeavor. But this isn't saying that other possibilities don't exist! If your community is without an emergency communications vehicle, look around—the solution may be sitting in your driveway.

Although a station wagon is ideal for such a program, just about any U.S.-made automobile can be adapted, providing it's not so old that it uses 6-volts DC. If you or other club members are reluctant to adapt the family buggy, check the used car lots. Our bet is that there's enough reserve in the club treasury not only to buy a reliable Detroit beast, but also to equip it for all contingencies. Bear in mind that maintenance is practically nil, since an emergency vehicle will probably not even come close to the same workouts that even a second family car receives weekly. Hence it will "live longer" and for less.

Basic advantages to having a "club car" as opposed to someone's family auto is that it can be taken out in all kinds of weather, equipped with heavy-duty snow tires the year round (for mud, gravel, etc.) and even painted an "off" color for distinctive recognition as an emergency vehicle. Although most localities prohibit solid red on a car not officially designed as law-enforcement, international orange is both striking and functional. Also off-grays can be considered. But best of all, you can rig up as many radios as you like without hearing about it from the XYLs.

Although a limited-channel rig can be used (insuring that

you have transmit and receive capability at least on Channels 9 and 11), a 23-channel CB is unquestionably the best suited for emergency vehicle use. If it's a tube set, make sure you have on hand at all times a carefully-packed spare set of tubes. Don't scrimp on your transceiver, but get the best your finances permit. It serves as the basic of your efforts.

When the car is in a fixed position, leave the hood open in a shaded area with the motor running whenever you're setting up shop. Protracted idling is hard enough on a car without a CB set draining off voltage. Pamper the battery, replacing it when necessary, and checking it on every trip for proper fluid level. Likewise, check the generator to insure that it charges properly.

Small gas-driven generators can be life-savers when operation for more than 10 or 15 minutes is contemplated, particularly during protracted emergencies in inclement weather. A warning, though: Generators like auto engines must be shielded, since they are just as prone to generate ignition noise as car engines—if not more so! If the generator has been rented, place it as far as possible from the CB rig, using heavy extension cords built for carrying 117v AC outdoors for power tools. The farther the generator from your car, the less the ignition noise will be. Additionally, generators aren't the quietest things in the world to start with.

Although conventional mobile CB antennas are convenient, they are sorely lacking when contrasted by their base-station "big brothers." For this reason, many CB clubs which own emergency vans or communications cars find it advantageous to rig up a ground-plane configuration which can be assembled and on the air in a matter of minutes. The way it works is unbelievably simple. A half-inch wooden plank measuring about 6 x 24" is used to support the antenna. At one end of the board a pipe flange is secured. Next, a hardware-store-variety 6" threaded pipe is screwed into the flange, allowing the hollow of the pipe to serve as a flag-pole-type support for the antenna mast. That's all there is to it. To use it, merely drive one wheel of the car over the other end of the plank. Now erect the antenna on a section of aluminum TV masting, guy it three ways to nearby trees, post, etc., and run the coax feedline to the CB rig under the dash. Believe it or not, one person can get this antenna into the air, but it goes a lot faster with two.

Location will be determined, of course, by the emergency or accident, although there are a few tips worth remembering: Always seek out the highest possible point, even though this may necessitate your walking 60 feet to the scene. The added range of the set will be well worth the trip. If your communications center must be set up on a main artery such as a U.S. highway, pull as far off the road as possible without getting stuck in mud or slush. If signal levels appear weak, move the car ahead or backwards a few feet. Often "dead spots" exist that can be circumvented quite satisfactorily in this manner. If it is known that certain routes or sections of roadway are trouble spots, take on-the-air trips over those roadways during leisure hours to familiarize yourself with propagational characteristics. Often you will find certain patches of road that are unusually suited for communications purposes (your contact may report, for example, a two "S"-unit improvement in signal strength). Make a mental note of these. Additionally, try various means of approach to known trouble spots. Should a serious accident occur, you may find the main arteries blocked or impassable. Prior knowledge of a backwoods shortcut may mean the difference between life or death for an injured motorist.

Don't forget to fill your communications vehicle with essentials. Food and a first aid kit are perhaps the most important, although elimination of any "must" items could lead to trouble. Always carry a detailed roadmap of the area of operations (Hagstrom's County Maps, showing off-beat dirt roads, are perfect). If intelligently followed through, you will wind up with a vehicular emergency system equal to (and sometimes better than) the best $20,000 communications vans in use today. This junior version, while perhaps not as elaborate, can be many times more functional in actual use providing it is kept in top-notch condition, and all operators know what to do in time of emergency. Coordination with the kind of base monitoring stations previously described is mandatory.

THE SIGN MARKER

A monitor marker sign on the outskirts of your town, village, or whatever, is your notification to passing CBers that full motorist services are available at all times. This means

Although this photo is actually of a mobile van used for promotional and demonstrational purposes by Aircraft Radio Corporation, Boonton, New Jersey, it illustrates what many CB emergency vans look like. Notice the gasoline driven generator at far left. This conserves the van's battery and assures full 117 VAC for all radio gear. The antenna, a 3-elementagi antenna, is mounted 20 feet up on the telescoping mast. The mast is simply turned by hand to rotate the array. Many communities and CB associations have had fund-raising campaigns to pay for such emergency vehicles.

that he can put in a call on Channel 11, for example, and ask about motel availabilities, eating places, as well as emergency aid. Should he require a tow-truck, he'll know that Channel 11 will get it for him. Likewise, should he spot an accident not yet attended to by authorities, he need only spot your roadsign, switch to the appropriate channel, and aid will be on its way.

To the non-CBer traveling the roadway, the sign is just another civic reminder that CBers are ready, willing, and able to offer their services. More and more publicity is being afforded citizens radio everyday, and bits of information relating to CB are beginning to filter down to the average motorist. Recent GM DAIR program publicity, coupled with Ford's Radio Road Alert program, and Project HELP, the CB emergency assistance program is becoming a national phenomena, and informed drivers are now aware that some kind of a race is on for two-way radio in the car. Even Motorola's CRW (Community Radio Watch) has alerted many citizens that mobile radio has definite value, even in combating the growth of crime. Hence, your roadsign will trigger the memories of more passing drivers than you'd imagine, and many will

Project HELP is discussed here at a New York editorial luncheon among influential members of N.Y. Electronics Press Club. Gentleman seated just right of center (facing camera) with beard is Tom Kneitel, KQD-4552, well-known editor of S9/The Citizens Band Journal.

take the trouble to inquire further when they reach their destination. Many CB clubs inscribe the name of their organization on the bottom of their alerting signs and make literature relating to CB available at truck stops and motels in the general vicinity. At any rate, everyone comes out ahead when some CB group goes to the trouble of erecting roadsigns. But this doesn't mean some quick-slap chalked affair stapled to a tree trunk!

Right off the bat, you will get yourself involved in a lot of red tape. You must contact local authorities if you don't want your sign ripped down for a local ordinance violation. Yet this isn't as foreboding as it might at first seem, since frequently these people will help rather than hinder your efforts by suggesting better locations, sign painters, sources of free materials, etc. First off, look to the local highway or road department, or the town's police or sheriff. If you contemplate erecting signs within your local township, there's no need for bringing in the state highway authorities.

Yes, there are numerous laws governing such signs, although such a worthy cause as road service alerting through a bonafide CB group does much to cut the red tape. Most states, for example, turn thumbs down on diamond or circular-shaped signs because these types are generally used to indicate highway instructions. Check out these laws beforehand, and you'll have smooth going of it.

Begin with a square or rectangular sign using a white background with red, blue, or black lettering, making the channel number as large as possible. To be easily seen, your sign should be at least two feet high by two feet wide, with lettering silk-screened or baked-on with enamel. Hand-painted wood signs do not hold up. Use instead 24 gauge solid steel, supported by a treated metal or wooden post at least nine feet long. The post should be cemented into a large pail and sunk a minimum of three feet into the ground. At this point, a good hunk of publicity can be captured, if everyone comes formally-dressed, an official is present, and all the photographers have been brought to the scene.

If no official newspaper coverage can be obtained in this manner, take a picture of the club president tapping the last shovelful of dirt over the post's base with plenty of club officers around (try to get your emergency vehicle into the picture). Write up a simple release, and you're practically assured of inclusion in tomorrow's edition. Don't neglect radio and TV publicity either. Remember that your club efforts help to assist <u>all</u> motorists, not just CBers.

LET EVERYONE INTO THE ACT

As indicated earlier, effective base or mobile communication and monitoring cannot be carried out alone but must be the culmination of a coordinated effort involving the very best CBers your locale and acquaintances have to offer. Every effort must be made to insure that the maximum number of serious-minded licensed CBers take part. In this way, the most coverage can be extended and result in the most service to the needy. REACT affiliation is of prime importance. Local CB club scheduling is a must. Twenty-four hour coverage must be maintained if the service is to be used effectively. Among your members there must be at least a few who operate the "graveyard shift" and consequently willing to fill in during the wee hours.

We can't stress too much the value of keeping an up-to-date log, recording every pertinent transmission along with detailed accounts of incidents where service has been rendered. Often this information is the only record made of road accidents serious enough to require assistance, but not so severe as to necessitate hospitalization or highway patrol inquiries. In any case, all participating persons, whether base monitors or mobile emergency operators, must keep accurate, clearly-written logs, a copy of which should be forwarded periodically to the club secretary.

Getting an "in" with the local police department communications people is probably the most rewarding contact you can establish. The Citizens Band Operating Committee of Elkhart, Indiana, for example, has worked hand-in-hand with the Elkhart Police Department even to the extent of setting up CB equipment at police HQ! The team effort has resulted in expanded aid and services during tornado conditions, flooding, etc. Besides helping established law-enforcement officers on search and rescue missions, organizations like the Operating Committee gain publicity and local recognition —an essential ingredient if full service capability is to be realized.

Appendix
RULES AND REGULATIONS
Part 95

Citizens Radio Service
Contents

Subpart A—General

Sec.
- 95.1 Basis and purpose.
- 95.3 Definitions.
- 95.5 Policy governing the assignment of frequencies.
- 95.6 **Types of operation authorized.**
- 95.7 General citizenship restrictions.

Subpart B—Applications and Licenses

- 95.11 Station authorization required.
- 95.13 Eligibility for station license.
- 95.15 Filing of applications.
- 95.17 Who may sign applications.
- 95.19 Standard forms to be used.
- 95.25 Amendment or dismissal of application.
- 95.27 Transfer of license prohibited.
- 95.29 Defective applications.
- 95.31 Partial grant.
- 95.33 License term.
- 95.35 Changes in transmitters and authorized stations.
- 95.37 Limitations on antenna structures.

Subpart C—Technical Regulations

- 95.41 Frequencies available.
- **95.42 Special provisions.**
- 95.43 Transmitter power.

95.44	External radio frequency power amplifiers prohibited.
95.45	Frequency tolerance.
95.47	Types of emission.
95.49	Emission limitations.
95.51	Modulation requirements.
95.53	Compliance with technical requirements.
95.55	Acceptability of transmitters for licensing.
95.57	Procedure for type acceptance of equipment.
95.58	Additional requirements for type acceptance.
95.59	Submission of noncrystal controlled Class C station transmitters for type approval.
95.61	Type approval of receiver-transmitter combinations.
95.63	Minimum equipment specifications.
95.65	Test procedure.
95.67	Certificate of type approval.

Subpart D—Station Operating Requirements

Sec.

95.83	Prohibited uses.
95.85	Emergency and assistance to motorist use.
95.87	Operation by, or on behalf of, persons other than the licensee.
95.89	Telephone answering services.
95.91	Duration of transmissions.
95.93	Tests and adjustments.
95.95	Station identification.
95.97	Operator license requirements.
95.101	Posting station license and transmitter identification cards or plates.
95.103	Inspection of stations and station records.
95.105	Current copy of rules required.
95.107	Inspection and maintenance of tower marking and lighting, and associated control equipment.
95.111	Recording of tower light inspections.
95.113	Answers to notices of violations.
95.115	False signals.
95.117	Station location.
95.119	Control points, dispatch points, and remote control.
95.121	Civil defense communications.

Subpart E—Operation of Citizens Radio Stations in the United States by Canadians

95.131	Basis, purpose and scope.
95.133	Permit required.
95.135	Application for permit.

95.137 Issuance of permit.
95.139 Modification or cancellation of permit.
95.141 Possession of permit.
95.143 Knowledge of rules required.
95.145 Operating conditions.
95.147 Station identification.

AUTHORITY: §§ 95.1 to 95.147 issued under secs. 4, 303, 48 Stat. 1066, 1082, as amended; 47 U.S.C. 154, 303. Interpret or apply 48 Stat. 1064–1068, 1081–1105, as amended; 47 U.S.C. Sub-chap. I, III–VI.

SUBPART A—GENERAL

§ 95.1 Basis and purpose.

The rules and regulations set forth in this part are issued pursuant to the provisions of Title III of the Communications Act of 1934, as amended, which vests authority in the Federal Communications Commission to regulate radio transmissions and to issue licenses for radio stations. These rules are designed to provide for private short-distance radiocommunications service for the business or personal activities of licensees, for radio signaling, for the control of remote objects or devices by means of radio; all to the extent that these uses are not specifically prohibited in this part. They also provide for procedures whereby manufacturers of radio equipment to be used or operated in the Citizens Radio Service may obtain type acceptance and/or type approval of such equipment as may be appropriate.

§ 95.3 Definitions.

For the purpose of this part, the following definitions shall be applicable. For other definitions, refer to Part 2 of this chapter.

(a) Definitions of services.

Citizens Radio Service. A radiocommunications service of fixed, land, and mobile stations intended for short-distance personal or business radiocommunications, radio signaling, and control of remote objects or devices by radio; all to the extent that these uses are not specifically prohibited in this part.

Fixed service. A service of radiocommunication between specified fixed points.

Mobile service. A service of radiocommunication between mobile and land stations or between mobile stations.

(b) Definitions of stations.

Base station. A land station in the land mobile service carrying on a service with land mobile stations.

Class A station. A station in the Citizens Radio Service licensed to be operated on an assigned frequency in the 460–470 MHz band with a transmitter output power of not more than 50 watts.

Class B station. (All operations terminated as of November 1, 1971.)

Class C station. A station in the Citizens Radio Service licensed to be operated on an authorized frequency in the 26.96–27.23 MHz band, or on the frequency 27.255 MHz, for the control of remote objects or devices by radio, or for the remote actuation of devices which are used solely as a means of attracting attention, or on an authorized frequency in the 72–76 MHz band for the radio control of models used for hobby purposes only.

Class D station. A station in the Citizens Radio Service licensed to be operated for radiotelephony, only, on authorized frequencies in the 26.96 MHz to 27.41 MHz band.

Fixed station. A station in the fixed service.

Land station. A station in the mobile service not intended for operation while in motion. (Of the various types of land stations, only the base station is pertinent to this part.)

Mobile station. A station in the mobile service intended to be used while in motion or during halts at unspecified points. (For the purposes of this part, the term includes hand-carried and pack-carried units.)

(c) Miscellaneous definitions.

Antenna structures. The term "antenna structures" includes the radiating system, its supporting structures and any appurtenances mounted thereon.

Assigned frequency. The frequency appearing on a station authorization from which the carrier frequency may deviate by an amount not to exceed that permitted by the frequency tolerance.

Authorized bandwidth. The maximum permissible bandwidth for the particular emission used. This shall be the occupied bandwidth or necessary bandwidth, whichever is greater.

Carrier power. The average power at the output terminals of a transmitter (other than a transmitter having a suppressed, reduced or controlled carrier) during one radio frequency cycle under conditions of no modulation.

Control point. A control point is an operating position which is under the control and supervision of the licensee, at which a person immediately responsible for the proper operation of the transmitter is stationed, and at which adequate means are available to aurally monitor all transmissions and to render the transmitter inoperative.

Dispatch point. A dispatch point is any position from which messages may be transmitted under the supervision of the person at a control point.

Double sideband emission. An emission in which both upper and lower sidebands resulting from the modulation of a particular carrier are transmitted. The carrier, or a portion thereof, also may be present in the emission.

External radio frequency power amplifiers. As defined in § 2.815(a) and as used in this part, an external radio frequency power amplifier is any device which, (1) when used in conjunction with a radio transmitter as a signal source is capable of amplification of that signal, and (2) is not an integral part of a radio transmitter as manufactured.

Harmful interference. Any emission, radiation or induction which endangers the functioning of a radionavigation service or other safety service or seriously degrades, obstructs or repeatedly interrupts a radiocommunication service operating in accordance with applicable laws, treaties, and regulations.

Man-made structure. Any construction other than a tower, mast or pole.

Mean power. The power at the output terminals of a transmitter during normal operation, averaged over a time sufficiently long compared with the period of the lowest frequency encountered in the modulation. A time of $1/10$ second during which the mean power is greatest will be selected normally.

Necessary bandwidth. For a given class of emission, the minimum value of the occupied bandwidth sufficient to ensure the transmission of information at the rate and with the quality required for the system employed, under specified conditions. Emissions useful for the good functioning of the receiving equipment, as for example, the emission corresponding to the carrier of reduced carrier systems, shall be included in the necessary bandwidth.

Occupied bandwidth. The frequency bandwidth such that, below its lower and above its upper frequency limits, the mean powers radiated are each equal to 0.5% of the total mean power radiated by a given emission.

Omnidirectional antenna. An antenna designed so the maximum radiation in any horizontal direction is within 3 dB of the minimum radiation in any horizontal direction.

Peak envelope power. The average power at the output terminals of a transmitter during one radio frequency cycle at the highest crest of the modulation envelope, taken under conditions of normal operation.

Person. The term "person" includes an individual, partnership, association, joint-stock company, trust or corporation.

Remote control. The term "remote control" when applied to the use or operation of a citizens radio station means control of the transmitting equipment of that station from any place other than the location of the transmitting equipment, except that direct mechanical control or direct electrical control by wired connections of transmitting equipment from some other point on the same premises, craft or vehicle shall not be considered to be remote control.

Single sideband emission. An emission in which only one sideband is transmitted. The carrier, or a portion thereof, also may be present in the emission.

Station authorization. Any construction permit, license, or special temporary authorization issued by the Commission.

§ 95.5 Policy governing the assignment of frequencies.

(a) The frequencies which may be assigned to Class A stations in the Citizens Radio Service, and the frequencies which are available for use by Class C or Class D stations are listed in Subpart C of this part. Each frequency available for assignment to, or use by, stations in this service is available on a shared basis only, and will not be assigned for the exclusive use of any one applicant; however, the use of a particular frequency may be restricted to (or in) one or more specified geographical areas.

(b) In no case will more than one frequency be assigned to Class A stations for the use of a single applicant in any given area until it has been demonstrated conclusively to the Commission that the assignment of an additional frequency is essential to the operation proposed.

(c) All applicants and licensees in this service shall cooperate in the selection and use of the frequencies assigned or authorized, in order to minimize interference and thereby obtain the most effective use of the authorized facilities.

(d) Simultaneous operation on more than one frequency in the 72–76 MHz band by a transmitter or transmitters of a single licensee is prohibited whenever such operation will cause harmful interference to the operation of other licensees in this service.

§ 95.6 Types of operation authorized.

(a) Class A stations may be authorized as mobile stations, as base stations, as fixed stations, or as base or fixed stations to be operated at unspecified or temporary locations.

(b) Class C and Class D stations are authorized as mobile stations only; however, they may be operated at fixed locations in accordance with other provisions of this part.

§ 95.7 General citizenship restrictions.

A station license may not be granted to or held by:
 (a) Any alien or the representative of any alien;
 (b) Any foreign government or the representative thereof;
 (c) Any corporation organized under the laws of any foreign government;
 (d) Any corporation of which any officer or director is an alien;
 (e) Any corporation of which more than one-fifth of the capital stock is owned of record or voted by: Aliens or their representatives; a foreign government or representative thereof; or any corporation organized under the laws of a foreign country;
 (f) Any corporation directly or indirectly controlled by any other corporation of which any officer or more than one-fourth of the directors are aliens, if the Commission finds that the public interest will be served by the refusal or revocation of such license; or
 (g) Any corporation directly or indirectly controlled by any other corporation of which more than one-fourth of the capital stock is owned of record or voted by: Aliens or their representatives; a foreign government or representatives thereof; or any corporation organized under the laws of a foreign government, if the Commission finds that the public interest will be served by the refusal or revocation of such license.

SUBPART B—APPLICATIONS AND LICENSES

§ 95.11 Station authorization required.

No radio station shall be operated in the Citizens Radio Service except under and in accordance with an authorization granted by the Federal Communications Commission.

§ 95.13 Eligibility for station license.

(a) Subject to the general restrictions of § 95.7, any person is eligible to hold an authorization to operate a station in the Citizens Radio Service: *Provided*, That if an applicant for a Class A or Class D station authorization is an individual or partnership, such individual or each partner is eighteen or more years of age; or if an applicant for a Class C station authorization is an individual or partnership, such individual or each partner is twelve or more years of age. An unincorporated association, when licensed under the provisions of this paragraph, may upon specific prior approval of the Commission provide radiocommunications for its members.

NOTE: While the basis of eligibility in this service includes any state, territorial, or local governmental entity, or any agency operating by the authority of such governmental entity, including any duly authorized state, territorial, or local civil defense agency, it should be noted that the frequencies available to stations in this service are shared without distinction between all licensees and that no protection is afforded to the communications of any station in this service from interference which may be caused by the authorized operation of other licensed stations.

(b) [Reserved]

(c) No person shall hold more than one Class C and one Class D station license.

§ 95.14 Mailing address furnished by licensee.

Except for applications submitted by Canadian citizens pursuant to agreement between the United States and Canada (TIAS No. 2508 and No. 6931), each application shall set forth and each licensee shall furnish the Commission with an address in the United States to be used by the Commission in serving documents or directing correspondence to that licensee. Unless any licensee advises the Commission to the contrary, the address contained in the licensee's most recent application will be used by the Commission for this purpose.

§ 95.15 Filing of applications.

(a) To assure that necessary information is supplied in a consistent manner by all persons, standard forms are prescribed for use in connection with the majority of applications and reports submitted for Commission consideration. Standard numbered forms applicable to the Citizens Radio Service are discussed in § 95.19 and may be obtained from the Washington, D.C., 20554, office of the Commission, or from any of its engineering field offices.

(b) All formal applications for Class C or Class D new, modified, or renewal station authorizations shall be submitted to the Commission's office at 334 York Street, Gettysburg, Pa. 17325. Applications for Class A station authorizations, applications for consent to transfer of control of a corporation holding any citizens radio station authorization, requests for special temporary authority or other special requests, and correspondence relating to an application for any class citizens radio station authorization shall be submitted to the Commission's Office at Washington, D.C. 20554, and should be directed to the attention of the Secretary. Beginning January 1, 1973, applicants for Class A stations in the Chicago Regional Area, defined in § 95.19, shall submit their applications to the Commission's Chicago Regional Office. The address of the Regional Office will be announced at a later date. Applications involving Class A or Class D station equipment which is neither type approved nor crystal controlled, whether of commercial or home construction, shall be accompanied by supplemental data describing in detail the design and construction of the transmitter and methods employed in testing it to determine compliance with the technical requirements set forth in Subpart C of this part.

(c) Unless otherwise specified, an application shall be filed at least 60 days prior to the date on which it is desired that Commission action thereon be completed. In any case where the applicant has made timely and sufficient application for renewal of license, in accordance with the Commission's rules, no license with reference to any activity of a continuing nature shall expire until such application shall have been finally determined.

(d) Failure on the part of the applicant to provide all the information required by the application form, or to supply the necessary exhibits or supplementary statements may constitute a defect in the application.

(e) Applicants proposing to construct a radio station on a site located on land under the jurisdiction of the U.S. Forest Service, U.S. Department of Agriculture, or the Bureau of Land Management, US. Department of the Interior, must supply the information and must follow the procedure prescribed by § 1.70 of this chapter.

§ 95.17 Who may sign applications.

(a) Except as provided in paragraph (b) of this section, applications, amendments thereto, and related statements of fact required by the Commission shall be personally signed by the applicant, if the applicant is an individual; by one of the partners, if the applicant is a partnership; by an officer, if the applicant is a corporation; or by a member who is an officer, if the applicant is an unincorporated association. Applications, amendments, and related statements of fact filed on behalf of eligible government entities, such as states and territories of the United States and political subdivisions thereof, the District of Columbia, and units of local government, including incorporated municipalities, shall be signed by such duly elected or appointed officials as may be competent to do so under the laws of the applicable jurisdiction.

(b) Applications, amendments thereto, and related statements of fact required by the Commission may be signed by the applicant's attorney in case of the applicant's physical disability or of his absence from the United States. The attorney shall in that event separately set forth the reason why the application is not signed by the applicant. In addition, if any matter is stated on the basis of the attorney's belief only (rather than his knowledge), he shall separately set forth his reasons for believing that such statements are true.

(c) Only the original of applications, amendments, or related statements of fact need be signed; copies may be conformed.

(d) Applications, amendments, and related statements of fact need not be signed under oath. Willful false statements made therein, however, are punishable by fine and imprisonment. U.S. Code, Title 18, section 1001, and by appropriate administrative sanctions, including revocation of station license pursuant to section 312(a)(1) of the Communications Act of 1934, as amended.

§ 95.19 Standard forms to be used.

(a) *FCC Form 505, Application for Class C or D Station License in the Citizens Radio Service.* This form shall be used when:

(1) Application is made for a new Class C or Class D authorization. A separate application shall be submitted for each proposed class of station.

(2) Application is made for modification of any existing Class C or Class D station authorization in those cases where prior Commission approval of certain changes is required (see § 95.35).

(3) Application is made for renewal of an existing Class C or Class D station authorization, or for reinstatement of such an expired authorization.

(b) *FCC Form 400, Application for Radio Station Authorization in the Safety and Special Radio Services.* Except as provided in paragraph (d) of this section, this form shall be used when:

(1) Application is made for a new Class A base station or fixed station authorization. Separate applications shall be submitted for each proposed base or fixed station at different fixed locations; however, all equipment intended to be operated at a single fixed location is considered to be one station which may, if necessary, be classed as both a base station and a fixed station.

(2) Application is made for a new Class A station authorization for any required number of mobile units (including hand-carried and pack-carried units) to be operated as a group in a single radiocommunication system in a particular area. An application for Class A mobile station authorization may be combined with the application for a single Class A base station au-

thorization when such mobile units are to be operated with that base station only.

(3) Application is made for station license of any Class A base station or fixed station upon completion of construction or installation in accordance with the terms and conditions set forth in any construction permit required to be issued for that station, or application for extension of time within which to construct such a station.

(4) Application is made for modification of any existing Class A station authorization in those cases where prior Commission approval of certain changes is required (see § 95.35).

(5) Application is made for renewal of an existing Class A station authorization, or for reinstatement of such an expired authorization.

(6) Each applicant in the Safety and Special Radio Services (1) for modification of a station license involving a site change or a substantial increase in tower height or (2) for a license for a new station must, before commencing construction, supply the environmental information, where required, and must follow the procedure prescribed by Subpart I of Part 1 of this chapter (§§ 1.1301 through 1.1319) unless Commission action authorizing such construction would be a minor action with the meaning of Subpart I of Part 1.

(7) Application is made for an authorization for a new Class A base or fixed station to be operated at unspecified or temporary locations. When one or more individual transmitters are each intended to be operated as a base station or as a fixed station at unspecified or temporary locations for indeterminate periods, such transmitters may be considered to comprise a single station intended to be operated at temporary locations. The application shall specify the general geographic area within which the operation will be confined. Sufficient data must be submitted to show the need for the proposed area of operation.

(c) *FCC Form 703, Application for Consent to Transfer of Control of Corporation Holding Construc-*

tion Permit or Station License. This form shall be used when application is made for consent to transfer control of a corporation holding any citizens radio station authorization.

(d) Beginning April 1, 1972, FCC Form 425 shall be used in lieu of FCC Form 400, applicants for Class A stations located in the Chicago Regional Area defined to consist of the counties listed below:

ILLINOIS

1. Boone.
2. Bureau.
3. Carroll.
4. Champaign.
5. Christian.
6. Clark.
7. Coles.
8. Cook.
9. Cumberland.
10. De Kalb.
11. De Witt.
12. Douglas.
13. Du Page.
14. Edgar.
15. Ford.
16. Fulton.
17. Grundy.
18. Henry.
19. Iroquois.
20. Jo Daviess.
21. Kane.
22. Kankakee.
23. Kendall.
24. Knox.
25. Lake.
26. La Salle.
27. Lee.
28. Livingston.
29. Logan.
30. Macon.
31. Marshall.
32. Mason.
33. McHenry.
34. McLean.
35. Menard.
36. Mercer.
37. Moultrie.
38. Ogle.
39. Peoria.
40. Piatt.
41. Putnam.
42. Rock Island.
43. Sangamon.
44. Shelby.
45. Stark.
46. Stephenson.
47. Tazewell.
48. Vermilion.
49. Warren.
50. Whiteside.
51. Will.
52. Winnebago.
53. Woodford.

INDIANA

1. Adams.
2. Allen.
3. Benton.
4. Blackford.
5. Boone.
6. Carroll.
7. Cass.
8. Clay.
9. Clinton.
10. De Kalb.
11. Delaware.
12. Elkhart.
13. Fountain.
14. Fulton.
15. Grant.
16. Hamilton.
17. Hancock.
18. Hendricks.
19. Henry.
20. Howard.
21. Huntington.
22. Jasper.

INDIANA—Continued

23. Jay.
24. Kosciusko.
25. Lake.
26. Lagrange.
27. La Porte.
28. Madison.
29. Marion.
30. Marshall.
31. Miami.
32. Montgomery.
33. Morgan.
34. Newton.
35. Noble.
36. Owen.
37. Parke.
38. Porter.
39. Pulaski.
40. Putnam.
41. Randolph.
42. St. Joseph.
43. Starke.
44. Steuben.
45. Tippecanoe.
46. Tipton.
47. Vermilion.
48. Vigo.
49. Wabash.
50. Warren.
51. Wells.
52. White.
53. Whitley.

IOWA

1. Cedar.
2. Clinton.
3. Dubuque.
4. Jackson.
5. Jones.
6. Muscatine.
7. Scott.

MICHIGAN

1. Allegan.
2. Barry.
3. Berrien.
4. Branch.
5. Calhoun.
6. Cass.
7. Clinton.
8. Eaton.
9. Hillsdale.
10. Ingham.
11. Ionia.
12. Jackson.
13. Kalamazoo.
14. Kent.
15. Lake.
16. Mason.
17. Mecosta.
18. Montcalm.
19. Muskegon.
20. Newaygo.
21. Oceana.
22. Ottawa.
23. St. Joseph.
24. Van Buren.

OHIO

1. Defiance.
2. Mercer.
3. Paulding.
4. Van Wert.
5. Williams.

WISCONSIN

1. Adams.
2. Brown.
3. Calumet.
4. Columbia.
5. Dane.
6. Dodge.
7. Door.
8. Fond du Lac.
9. Grant.
10. Green.

11. Green Lake.
12. Iowa.
13. Jefferson.
14. Juneau.
15. Kenosha.
16. Kewaunee.
17. Lafayette.
18. Manitowoc.
19. Marquette.
20. Milwaukee.
21. Outagamie.
22. Ozaukee.
23. Racine.
24. Richland.
25. Rock.
26. Sauk.
27. Sheboygan.
28. Walworth.
29. Washington.
30. Waukesha.
31. Waupaca.
32. Waushara.
33. Winnebago.

§ 95.25 Amendment or dismissal of application.

(a) Any application may be amended upon request of the applicant as a matter of right prior to the time the application is granted or designated for hearing. Each amendment to an application shall be signed and submitted in the same manner and with the same number of copies as required for the original application.

(b) Any application may, upon written request signed by the applicant or his attorney, be dismissed without prejudice as a matter of right prior to the time the application is granted or designated for hearing.

§ 95.27 Transfer of license prohibited.

A station authorization in the Citizens Radio Service may not be transferred or assigned. In lieu of such transfer or assignment, an application for new station authorization shall be filed in each case, and the previous authorization shall be forwarded to the Commission for cancellation.

§ 95.29 Defective applications.

(a) If an applicant is requested by the Commission to file any documents or information not included in the prescribed application form, a failure to comply with such request will constitute a defect in the application.

(b) When an application is considered to be incomplete or defective, such application will be returned to the applicant, unless the Commission may otherwise direct. The reason for return of the applications will be indicated, and if appropriate, necessary additions or corrections will be suggested.

§ 95.31 Partial grant.

Where the Commission, without a hearing, grants an application in part, or with any privileges, terms, or conditions other than those requested, the action of the Commission shall be considered as a grant of such application unless the applicant shall, within 30 days from the date on which such grant is made, or from its effective date if a later date is specified, file with the Commission a written rejection of the grant as made. Upon receipt of such rejection, the Commission will vacate its original action upon the application and, if appropriate, set the application for hearing.

§ 95.33 License term.

Licenses for stations in the Citizens Radio Service will normally be issued for a term of 5 years from the date of original issuance, major modification, or renewal.

§ 95.35 Changes in transmitters and authorized stations.

Authority for certain changes in transmitters and authorized stations must be obtained from the Commission before the changes are made, while other changes do not require prior Commission approval. The following paragraphs of this section describe the conditions under which prior Commission approval is or is not necessary.

(a) Proposed changes which will result in operation inconsistent with any of the terms of the current authorization require that an application for modification of license be submitted to the Commission. Application for modification shall be submitted in the same manner as an application for a new station license, and the licensee shall forward his existing authorization to the Commission for cancellation immediately upon receipt of the superseding authorization. Any of the following changes to authorized stations may be made only upon approval by the Commission:

(1) Increase the overall number of transmitters authorized.

(2) Change the presently authorized location of a Class A fixed or base station or control point.

(3) Move, change the height of, or erect a Class A station antenna structure.

(4) Make any change in the type of emission or any increase in bandwidth of emission or power of a Class A station.

(5) Addition or deletion of control point(s) for an authorized transmitter of a Class A station.

(6) Change or increase the area of operation of a Class A mobile station or a Class A base or fixed station authorized to be operated at temporary locations.

(7) Change the operating frequency of a Class A station.

(b) When the name of a licensee is changed (without changes in the ownership, control, or corporate structure), or when the mailing address of the licensee is changed (without changing the authorized location of the base or fixed Class A station) a formal application for modification of the license is not required. However, the licensee shall notify the Commission promptly of these changes. The notice, which may be in letter form, shall contain the name and address of the licensee as they appear in the Commission's records, the new name and/or address, as the case may be, and the call signs and classes of all radio stations authorized to the licensee under this part. The notice concerning Class C or D radio stations shall be sent to Federal Communications Commission, Gettysburg, Pa. 17325, and a copy shall be maintained with the records of the station. The notice concerning Class A stations shall be sent to (1) Secretary, Federal Communications Commission, Washington, D.C. 20554, and (2) to Engineer in Charge of the Radio District in which the station is located, and a copy shall be maintained with the license of the station until a new license is issued.

(c) Proposed changes which will not depart from any of the terms of the outstanding authorization for the station may be made without prior Commission approval. Included in such changes is the substitution of transmitting equipment at any station, provided that the equipment employed is included in the Commission's

"Radio Equipment List," and is listed as acceptable for use in the appropriate class of station in this service. Provided it is crystal-controlled and otherwise complies with the power, frequency tolerance, emission and modulation percentage limitations prescribed, non-type accepted equipment may be substituted at:

(1) Class C stations operated on frequencies in the 26.99–27.26 MHz band;

(2) Class D stations until November 22, 1974.

(d) Transmitting equipment type accepted for use in Class D stations shall not be modified by the user. Changes which are specifically prohibited include:

(1) Internal or external connection or addition of any part, device or accessory not included by the manufacturer with the transmitter for its type acceptance. This shall not prohibit the external connection of antennas or antenna transmission lines, antenna switches, passive networks for coupling transmission lines or antennas to transmitters, or replacement of microphones.

(2) Modification in any way not specified by the transmitter manufacturer and not approved by the Commission.

(3) Replacement of any transmitter part by a part having different electrical characteristics and ratings from that replaced unless such part is specified as a replacement by the transmitter manufacturer.

(4) Substitution or addition of any transmitter oscillator crystal unless the crystal manufacturer or transmitter manufacturer has made an express determination that the crystal type, as installed in the specific transmitter type, will provide that transmitter type with the capability of operating within the frequency tolerance specified in Section 95.45(a).

(5) Addition or substitution of any component, crystal or combination of crystals, or any other alteration to enable transmission on any frequency not authorized for use by the licensee.

(e) Only the manufacturer of the particular unit of equipment type accepted for use in Class D stations may make the permissive changes allowed under the provisions of Part 2 of this chapter for type acceptance.

However, the manufacturer shall not make any of the following changes to the transmitter without prior written authorization from the Commission:

(1) Addition of any accessory or device not specified in the application for type acceptance and approved by the Commission in granting said type acceptance.

(2) Addition of any switch, control, or external connection.

(3) Modification to provide capability for an additional number of transmitting frequencies.

§ 95.37 Limitations on antenna structures.

(a) Except as provided in paragraph (b) of this section, an antenna for a Class A station which exceeds the following height limitations may not be erected or used unless notice has been filed with both the FAA on FAA Form 7460-1 and with the Commission on Form 714 or on the license application form, and prior approval by the Commission has been obtained for:

(1) Any construction or alteration of more than 200 feet in height above ground level at its site (§ 17.7(a) of this chapter).

(2) Any construction or alteration of greater height than an imaginary surface extending outward and upward at one of the following slopes (§ 17.7(b) of this chapter):

(i) 100 to 1 for a horizontal distance of 20,000 feet from the nearest point of the nearest runway of each airport with at least one runway more than 3,200 feet in length, excluding heliports, and seaplane bases without specified boundaries, if that airport is either listed in the Airport Directory of the current Airman's Information Manual or is operated by a Federal military agency.

(ii) 50 to 1 for a horizontal distance of 10,000 feet from the nearest point of the nearest runway of each airport with its longest runway no more than 3,200 feet in length, excluding heliports, and seaplane bases without specified boundaries, if that airport is either listed in the Airport Directory or is operated by a Federal military agency.

(iii) 25 to 1 for a horizontal distance of 5,000 feet from the nearest point of the nearest landing and take-off area of each heliport listed in the Airport Directory or operated by a Federal military agency.

(3) Any construction or alteration on any airport listed in the Airport Directory of the current Airman's Information Manual (§ 17.7(c) of this chapter).

(b) A notification to the Federal Aviation Administration is not required for any of the following construction or alteration of Class A station antenna structures.

(1) Any object that would be shielded by existing structures of a permanent and substantial character or by natural terrain or topographic features of equal or greater height, and would be located in the congested area of a city, town, or settlement where it is evident beyond all reasonable doubt that the structure so shielded will not adversely affect safety in air navigation. Applicants claiming such exemption shall submit a statement with their application to the Commission explaining the basis in detail for their finding (§ 17.14(a) of this chapter).

(2) Any antenna structure of 20 feet or less in height except one that would increase the height of another antenna structure (§17.14(b) of this chapter).

(1) All antennas, both receiving and transmitting, and supporting structures associated or used in conjunction with a Class C or D citizens radio station operated from a fixed location must comply with at least one of the following:

(1) The antenna and its supporting structure does not exceed 20 feet in height above ground level; or

(2) The antenna and its supporting structure does not exceed by more than 20 feet the height of any natural formation, tree or man-made structure on which it is mounted; or

NOTE: A man-made structure is any construction other than a tower, mast, or pole.

(3) The antenna is mounted on the transmitting antenna structure of another authorized radio station and exceeds neither 60 feet above ground level nor the height of the antenna supporting structure of the other station; or

(4) The antenna is mounted on and does not exceed the height of the antenna structure otherwise used solely for receiving purposes, which structure itself complies with subparagraph (1) or (2) of this paragraph.

(5) The antenna is omnidirectional and the highest point of the antenna and its supporting structure does not exceed 60 feet above ground level and the highest point also does not exceed one foot in height above the established airport elevation for each 100 feet of horizontal distance from the nearest point of the nearest airport runway.

NOTE: A work sheet will be made available upon request to assist in determining the maximum permissible height of an antenna structure.

(d) Class C stations operated on frequencies in the 72–76 MHz band shall employ a transmitting antenna which complies with all of the following:

(1) The gain of the antenna shall not exceed that of a half-wave dipole;

(2) The antenna shall be immediately attached to, and an integral part of, the transmitter; and

(3) Only vertical polarization shall be used.

(e) Further details as to whether an aeronautical study and/or obstruction marking and lighting may be required, and specifications for obstruction marking and lighting when required, may be obtained from Part 17 of this chapter, "Construction, Marking, and Lighting of Antenna Structures."

(f) Subpart I of Part 1 of this chapter contains procedures implementing the National Environmental Policy Act of 1969. Applications for authorization of the construction of certain classes of communications facilities defined as "major actions" in § 1.305 thereof, are required to be accompanied by specified statements. Generally these classes are:

(1) Antenna towers or supporting structures which exceed 300 feet in height and are not located in areas devoted to heavy industry or to agriculture.

(2) Communications facilities to be located in the following areas:

(i) Facilities which are to be located in an officially designated wilderness area or in an area whose designation as a wilderness is pending consideration;

(ii) Facilities which are to be located in an officially designated wildlife preserve or in an area whose designation as a wildlife preserve is pending consideration;

(iii) Facilities which will affect districts, sites, buildings, structures or objects, significant in American history, architecture, archaeology or culture, which are listed in the National Register of Historic Places or are eligible for listing (see 36 CFR 800.2 (d) and (f) and 800.10); and

(iv) Facilities to be located in areas which are recognized either nationally or locally for their special scenic or recreational value.

(3) Facilities whose construction will involve extensive change in surface features (e.g. wetland fill, deforestation or water diversion).

NOTE: The provisions of this paragraph do not include the mounting of FM, television or other antennas comparable thereto in size on an existing building or antenna tower. The use of existing routes, buildings and towers is an environmentally desirable alternative to the construction of new routes or towers and is encouraged.

If the required statements do not accompany the application, the pertinent facts may be brought to the attention of the Commission by any interested person during the course of the license term and considered de novo by the Commission.

SUBPART C—TECHNICAL REGULATIONS

§ 95.41 Frequencies available.

(a) Frequencies available for assignment to Class A stations:

(1) The following frequencies or frequency pairs are available primarily for assignment to base and mobile stations. They may also be assigned to fixed stations as follows:

(i) Fixed stations which are used to control base stations of a system may be assigned the frequency assigned to the mobile units associated with the base station. Such fixed stations shall comply with the following requirements if they are located within 75 miles of the center of urbanized areas of 200,000 or more population.

(*a*) If the station is used to control one or more base stations located within 45 degrees of azimuth, a directional antenna having a front-to-back ratio of at least 15 dB shall be used at the fixed station. For other situations where such a directional antenna cannot be used, a cardioid, bidirectional or omnidirectional antenna may be employed. Consistent with reasonable design, the antenna used must, in each case, produce a radiation pattern that provides only the coverage necessary to permit satisfactory control of each base station and limit radiation in other directions to the extent feasible.

(*b*) The strength of the signal of a fixed station controlling a single base station may not exceed the signal strength produced at the antenna terminal of the base receiver by a unit of the associated mobile station, by more than 6 dB. When the station controls more than one base station, the 6 dB control-to-mobile signal difference need be verified at only one of the base station sites. The measurement of the signal strength of the mobile unit must be made when such unit is transmitting from the control station location or, if that is not practical, from a location within one-fourth mile of the control station site.

(*c*) Each application for a control station to be authorized under the provisions of this paragraph shall be accompanied by a statement certifying that the output power of the proposed station transmitter will be adjusted to comply with the foregoing signal level limitation. Records of the measurements used to determine the signal ratio shall be kept with the station records and shall be made available for inspection by Commission personnel upon request.

(*d*) Urbanized areas of 200,000 or more population are defined in the U.S. Census of Population, 1960, Vol. 1, table 23, page 50. The centers of urbanized areas are determined from the Appendix, page 226 of the U.S. Commerce publication "Air Line Distance Between Cities in the United States."

(ii) Fixed stations, other than those used to control

base stations, which are located 75 or more miles from the center of an urbanized area of 200,000 or more population. The centers of urbanized areas of 200,000 or more population are listed on page 226 of the Appendix to the U.S. Department of Commerce publication "Air Line Distance Between Cities in the United States." When the fixed station is located 100 miles or less from the center of such an urbanized area, the power output may not exceed 15 watts. All fixed systems are limited to a maximum of two frequencies and must employ directional antennas with a front-to-back ratio of at least 15 dB. For two-frequency systems, separation between transmit-receive frequencies is 5 MHz.

Base and Mobile (MHz)	Mobile Only (MHz)
462.550	467.550
462.575	467.575
462.600	467.600
462.625	467.625
462.650	467.650
462.675	467.675
462.700	467.700
462.725	467.725

(2) Conditions governing the operation of stations authorized prior to March 18, 1968:

(i) All base and mobile stations authorized to operate on frequencies other than those listed in subparagraph (1) of this paragraph may continue to operate on those frequencies only until January 1, 1970.

(ii) Fixed stations located 100 or more miles from the center of any urbanized area of 200,000 or more population authorized to operate on frequencies other than those listed in subparagraph (1) of this paragraph will not have to change frequencies provided no interference is caused to the operation of stations in the land mobile service.

(iii) Fixed stations, other than those used to control base stations, located less than 100 miles (75 miles if the transmitter power output does not exceed 15 watts) from the center of any urbanized area of 200,000 or more population must discontinue operation by November 1, 1971. However, any operation after January 1, 1970, must be on frequencies listed in subparagraph (1) of this paragraph.

(iv) Fixed stations, located less than 100 miles from the center of any urbanized area of 200,000 or more population, which are used to control base stations and are authorized to operate on frequencies other than those listed in subparagraph (1) of this paragraph may continue to operate on those frequencies only until January 1, 1970.

(v) All fixed stations must comply with the applicable technical requirements of subparagraph (1) relating to antennas and radiated signal strength of this paragraph by November 1, 1971.

(vi) Notwithstanding the provisions of subdivisions (i) through (v) of this subparagraph, all stations authorized to operate on frequencies between 465.000 and 465.500 MHz and located within 75 miles of the center of the 20 largest urbanized areas of the United States, may continue to operate on these frequencies only until January 1, 1969. An extension to continue operation on such frequencies until January 1, 1970, may be granted to such station licensees on a case by case basis if the Commission finds that continued operation would not be inconsistent with planned usage of the particular frequency for police purposes. The 20 largest urbanized areas can be found in the U.S. Census of Population, 1960, vol. 1, table 23, page 50. The centers of urbanized areas are determined from the appendix, page 226, of the U.S. Commerce publication, "Air Line Distance Between Cities in the United States."

(b) [Reserved]

(c) Class C mobile stations may employ only amplitude tone modulation or on-off keying of the unmodulated carrier, on a shared basis with other stations in the Citizens Radio Service on the frequencies and under the conditions specified in the following tables:

(1) For the control of remote objects or devices by radio, or for the remote actuation of devices which are used solely as a means of attracting attention and subject to no protection from interference due to the

operation of industrial, scientific, or medical devices within the 26.96–27.28 MHz band, the following frequencies are available:

(MHz)	(MHz)	(MHz)
26.995	27.095	27.195
27.045	27.145	[1] 27.255

[1] The frequency 27.255 MHz also is shared with stations in other services.

(2) Subject to the conditions that interference will not be caused to the remote control of industrial equipment operating on the same or adjacent frequencies and to the reception of television transmissions on Channels 4 or 5; and that no protection will be afforded from interference due to the operation of fixed and mobile stations in other services assigned to the same or adjacent frequencies in the band, the following frequencies are available solely for the radio remote control of models used for hobby purposes:

(i) For the radio remote control of any model used for hobby purposes:

MHz	MHz	MHz
72.16	72.32	72.96

(ii) For the radio remote control of aircraft models only:

MHz	MHz	MHz
72.08	72.24	72.40
75.64		

(d) The frequencies listed in the following paragraphs are available for use by Class D stations and are subject to no protection from interference resulting from the operation of industrial, scientific, or medical devices in the 26.96 MHz to 27.28 MHz band.

(1) The following frequencies may be used for communications between Class D stations:

MHz	MHz	MHz	MHz
26.965	27.115	27.035	27.185
26.975	27.125	27.055	27.205
26.985	27.135	27.075	27.215
27.005	27.155	27.085	27.225
27.015	27.165	27.105	27.255
27.025	27.175		

(2) Effective January 1, 1977, the following frequencies may be used for communications between Class D stations:

MHz	MHz	MHz	MHz
26.965	27.105	27.225	27.325
26.965	27.115	27.235	27.335
26.985	27.125	27.245	27.345
27.005	27.135	27.255	27.355
27.015	27.155	27.265	27.365
27.025	27.165	27.275	27.375
27.035	27.175	27.285	27.385
27.055	27.185	27.295	27.395
27.075	27.205	27.305	27.405
27.085	27.215	27.315	

(3) The frequency 27.065 MHz shall be used solely for:
(i) Emergency communications involving the immediate safety of life of individuals or the immediate protection of property, or
(ii) Communications necessary to render assistance to a motorist.

NOTE:—A licensee, before using 27.065 MHz must make a determination that his communication is either or both (a) an emergency communication or (b) is necessary to render assistance to a motorist. To be an emergency communication, the message must have some direct relation to the immediate safety of life or immediate protection of property. If no immediate action is required, it is not an emergency. What may not be an emergency under one set of circumstances may be an emergency under different circumstances. There are many worthwhile public service communications that do not qualify as emergency communications. In the case of motorist assistance, the message must be necessary to assist a particular motorist and not, expect in a valid emergency, motorists in general. If the communications are to be lengthy, the exchange should be shifted to another frequency, if feasible, after contact is established. No nonemergency or nonmotorist assistance communications are permitted on 27.065 MHz even for the limited purpose of calling a licensee monitoring a frequency to ask him to switch to another frequency. Although 27.065 MHz may be used for marine emergencies, it should not be considered a substitute for the authorized marine distress system. The Coast Guard has stated it will not "participate directly in the Citizens Radio Service by fitting with and/or providing a watch on any Citizens Band Channel. (Coast Guard Commandant Instructions 2302.6)"

The following are examples of permitted and prohibited types of communications. They are guidelines and are not intended to be all inclusive.

Permitted	Example message
Yes	A tornado is sighted six miles north of town.
No	This is observation post number 10. No tornados sighted.
Yes	I am out of gas on Interstate 95.
No	I am out of gas in my driveway.
Yes	There is a four-car collision at Exit 10 on the Beltway. Send police and ambulance.
No	Traffic is moving smoothly on the Beltway.
Yes	Base to Unit 1. the Weather Bureau has just issued a thunderstorm warning. Bring the sailboat into port.
No	Attention all motorists. The Weather Bureau advises that the snow tomorrow will accumulate 4 to 6 inches.
Yes	There is a fire in the building on the corner of 6th and Main Streets.
No	This is Halloween patrol unit number 3. Everything is quiet here.

The following priorities should be observed in the use of 27.065 MHz:

1. Communications relating to an existing situation dangerous to life or property. i.e., fire, automobile accident.

2. Communications relating to a potentially hazardous situation, i.e., car stalled in a dangerous place, lost child, boat out of gas.

3. Road assistance to a disabled vehicle on the highway or street.

4. Road and street directions.

95.42 Special provisions.

Effective September 10, 1976 station authorizations for the use of frequencies between 26.96 MHz and 27.41 MHz will be issued only to applicants in the Citizens Radio Service. Any license in a radio service other than the Citizens Radio Service authorizing the use of frequencies between 26.96 MHz and 27.41 MHz shall remain valid until December 31, 1979.

§ 95.43 Transmitter power.

(a) Transmitter power is the power at the transmitter output terminals and delivered to the antenna, antenna transmission line, or any other impedance-matched, radio frequency load.

(1) For single sideband transmitters and other transmitters employing a reduced carrier, a suppressed carrier or a controlled carrier, used at Class D stations, transmitter power is the peak envelope power.

(2) For all transmitters other than those covered by paragraph (a)(1) of this section, the transmitter power is the carrier power.

(b) The transmitter power of a station shall not exceed the following values under any condition of modulation or other circumstances.

Class of station:	Transmitter power in watts
A	50
C—27.255 MHz	25
C—26.995–27.195 MHz	4
C—72–76 MHz	0.75
D—Carrier (where applicable)	4
D—Peak envelope power (where applicable)	12

§ 95.44 External radio frequency power amplifiers prohibited.

No external radio frequency power amplifier shall be used or attached, by connection, coupling attachment or in any other way at any Class D station.

NOTE: An external radio frequency power amplifier at a Class D station will be presumed to have been used where it is in the operator's possession or on his premises and there is extrinsic evidence of any operation of such Class D station in excess of power limitations provided under this rule part unless the operator of such equipment holds a station license in another radio service under which license the use of the said amplifier at its maximum rated output power is permitted.

§ 95.45 Frequency tolerance.

(a) Except as provided in paragraphs (b) and (c) of this section, the carrier frequency of a transmitter in this service shall be maintained within the following percentage of the authorized frequency:

Class of station	Frequency tolerance	
	Fixed and base	Mobile
A	0.00025	0.0005
C		.005
D		.005

(b) Transmitters used at Class C stations operating on authorized frequencies between 26.99 and 27.26 MHz with 2.5 watts or less mean output power, which are used solely for the control of remote objects or devices by radio (other than devices used solely as a means of attracting attention), are permitted a frequency tolerance of 0.01 percent.

(c) Class A stations operated at a fixed location used to control base stations, through use of a mobile only frequency, may operate with a frequency tolerance of 0.0005 percent.

§ 95.47 Types of emission.

(a) Except as provided in paragraph (e) of this section, Class A stations in this service will normally be authorized to transmit radiotelephony only. However, the use of tone signals or signaling devices solely to actuate receiver circuits, such as tone operated squelch or selective calling circuits, the primary function of which is to establish or establish and maintain voice communications, is permitted. The use of tone signals solely to attract attention is prohibited.

(b) [Reserved]

(c) Class C stations in this service are authorized to use amplitude tone modulation or on-off unmodulated carrier only, for the control of remote objects or devices by radio or for the remote actuation of devices which are used solely as a means of attracting attention. The transmission of any form of telegraphy, telephony or record communications by a Class C station is prohibited. Telemetering, except for the transmission of simple, short duration signals indicating the presence or absence of a condition or the occurrence of an event, is also prohibited.

(d) Transmitters used at Class D stations in this service are authorized to use amplitude voice modulation, either single or double sideband. Tone signals or signalling devices may be used only to actuate receiver circuits, such as tone operated squelch or selective calling circuits, the primary function of which is to establish or maintain voice communications. The use of any signals solely to attract attention or for the control of remote objects or devices is prohibited.

(e) Other types of emission not described in paragraph (a) of this section may be authorized for Class A citizens radio stations upon a showing of need therefor. An application requesting such authorization shall fully describe the emission desired, shall indicate the bandwidth required for satisfactory communication, and shall state the purpose for which such emission is required. For information regarding the classification of emissions and the calculation of bandwidth, reference should be made to Part 2 of this chapter.

§ 95.49 Emission limitations.

(a) Each authorization issued to a Class A citizens radio station will show, as a prefix to the classification of the authorized emission, a figure specifying the maximum bandwidth to be occupied by the emission.

(b) [Reserved]

(c) The authorized bandwidth of the emission of any transmitter employing amplitude modulation shall be 8 kHz for double sideband and 4 kHz for single sideband. The authorized bandwidth of the emission of any transmitter employing frequency or phase modulation (Class F2 or F3) shall be 20 kHz. The use of F2 and F3 emissions in the frequency band 26.96 MHz-27.41 MHz is not authorized.

(d) The mean power of emissions shall be attenuated below the mean power of the transmitter in accordance with the following schedule:

(1) When using emissions other than single sideband:

(i) On any frequency removed from the center of the authorized bandwidth by more than 50 percent up to and including 100 percent of the authorized bandwidth: at least 25 decibels;

(ii) On any frequency removed from the center of the authorized bandwidth by more than 100 percent up to and including 250 percent of the authorized bandwidth: At least 35 decibels;

(2) When using single sideband emissions:

(i) On any frequency removed from the center of the authorized bandwidth by more than 50 percent up to and including 150 percent of the authorized bandwidth: At least 25 decibels:

(ii) On any frequency removed from the center of the authorized bandwidth by more than 150 percent up to and including 250 percent of the authorized bandwidth: At least 35 decibels;

(3) On any frequency removed from the center of the authorized bandwidth by more than 250 percent of the authorized bandwidth: at least $43+10 \log_{10}$ (mean power in watts) decibels, for Class D transmitters type accepted before September 10, 1976 and all Class A transmitters.

(4) On any frequency removed from the center of the authorized bandwidth by more than 250 percent of the authorized bandwidth up to a frequency of twice the fundamental frequency; at least $53+10 \log_{10}$ (mean power in watts) decibels, for Class D transmitters type accepted after September 10, 1976.

(5) On any frequency twice or greater than twice the fundamental frequency: at least 60 decibels (mean power in watts) for Class D transmitters type accepted after September 10, 1976.

NOTE.—The requirements of paragraph (d) must be met both with and without connection of all attachments acceptable for use with such transmitters. External speakers, microphones, power cords, and antennas are among the devices included in this requirement. Additionally, if it is shown that a licensee causes interference to television reception because of insufficient harmonic attenuation, he may be required to insert a low pass filter between the transmitter RF output terminal and the antenna feedline.

(e) When an unauthorized emission results in harmful interference, the Commission may, in its discretion, require appropriate technical changes in equipment to alleviate the interference.

§ 95.51 **Modulation requirements.**

(a) When double sideband, amplitude modulation is used for telephony, the modulation percentage shall be sufficient to provide efficient communication and shall not exceed 100 percent.

(b) Each transmitter for use in Class D stations, other than single sideband, suppressed carrier, or controlled carrier, for which type acceptance is requested after May 24, 1974, having more than 2.5 watts maximum output power shall be equipped with a device which automatically prevents modulation in excess of 100 percent on positive and negative peaks.

(c) The maximum audio frequency required for satisfactory radiotelephone intelligibility for use in this service is considered to be 3000 Hz.

(d) Transmitters for use at Class A stations shall be provided with a device which automatically will prevent greater than normal audio level from causing modulation in excess of that specified in this subpart; *Provided, however,* That the requirements of this paragraph shall not apply to transmitters authorized at mobile stations and having an output power of 2.5 watts or less.

(e) Each transmitter of a Class A station which is equipped with a modulation limiter in accordance with the provisions of paragraph (d) of this section shall also be equipped with an audio low-pass filter. This audio low-pass filter shall be installed between the modulation limiter and the modulated stage and, at audio frequencies between 3 kHz and 20 kHz, shall have an attenuation greater than the attenuation at 1 kHz by at least:

$$60 \log_{10} (f/3) \text{ decibels}$$

where "f" is the audio frequency in kHz. At audio frequencies above 20 kHz, the attentuation shall be at least 50 decibels greater than the attenuation at 1 kHz.

(f) Simultaneous amplitude modulation and frequency or phase modulation of a transmitter is not authorized.

(g) The maximum frequency deviation of frequency modulated transmitters used at Class A stations shall not exceed ±5 kHz.

§ 95.53 **Compliance with technical requirements.**

(a) Upon receipt of notification from the Commission of a deviation from the technical requirements of the rules in this part, the radiations of the transmitter involved shall be suspended immediately, except for necessary tests and adjustments, and shall not be resumed until such deviation has been corrected.

(b) When any citizens radio station licensee receives a notice of violation indicating that the station has been operated contrary to any of the provisions contained in Subpart C of this part, or where it otherwise appears that operation of a station in this service may not be in accordance with applicable technical standards, the Commission may require the licensee to conduct such tests as may be necessary to determine whether the equipment is capable of meeting these standards and to make such adjustments as may be necessary to assure compliance therewith. A licensee who is notified that he is required to conduct such tests and/or make adjustments must, within the time limit specified in the notice, report to the Commission the results thereof.

(c) All tests and adjustments which may be required in accordance with paragraph (b) of this section shall be made by, or under the immediate supervision of, a person holding a first- or second-class commercial operator license, either radiotelephone or radio telegraph as may be appropriate for the type of emission employed. In each case, the report which is submitted to the Commission shall be signed by the licensed commercial operator. Such report shall describe the results of the tests and adjustments, the test equipment and procedures used, and shall state the type, class, and serial number of the operator's license. A copy of this report shall also be kept with the station records.

§ 95.55 **Acceptability of transmitters for licensing.**

Transmitters type approved or type accepted for use under this part are included in the Commission's Radio Equipment List. Copies of this list are available for public reference at the Commission's Washington, D.C., offices and field offices. The requirements for transmitters which may be operated under a license in this service are set forth in the following paragraphs.

(a) Class A stations: All transmitters shall be type accepted.

(b) Class C stations:

(1) Transmitters operated in the band 72–76 MHz shall be type accepted.

(2) All transmitters operated in the band 26.99–27.26 MHz shall be type approved, type accepted or crystal controlled.

(c) Class D Stations:

(1) All transmitters first licensed, or marketed as specified in § 2.805 of this chapter, prior to November 22, 1974, shall be type accepted or crystal controlled.

(2) All transmitters first licensed, or marketed as specified in § 2.803 of this chapter, on or after November 22, 1974, shall be type accepted.

(3) Effective November 23, 1978, all transmitters shall be type accepted.

(4) Priority to January 1, 1977 transmitters which are equipped to operate on any frequency not included in 95.41 (d) (1) may not be installed at, or used by, any Class D station unless there is a station license posted at the transmitter location, or a transmitter identification card (FCC Form 452-C) attached to the transmitter, which indicates that operation of the transmitter on such frequency has been authorized by the Commission.

(5) Effective January 1, 1977 transmitters which are equipped to operate on any frequency not included in 95.41 may not be installed at or used by any Class D station unless there is a station license posted at the transmitter location, or a transmitter identification card (FCC Form 452-C) attached to the transmitter, which indicates that operation of the transmitter on such frequency has been authorized by the Commission.

NOTE.—A "transmitter" is defined to include any radio frequency (RF) power amplifier.

(d) With the exception of equipment type approved for use at a Class C station, all transmitting equipment authorized in this service shall be crystal controlled.

(e) No controls, switches or other functions which can cause operation in violation of the technical regulations of this part shall be accessible from the operating panel or exterior to the cabinet enclosing a transmitter authorized in this service.

§ 95.57 **Procedure for type acceptance of equipment.**

(a) Any manufacturer of a transmitter built for use in this service, except noncrystal controlled transmitters for use at Class C stations, may request type acceptance for such transmitter in accordance with the type acceptance requirements of this part, following the type acceptance procedure set forth in Part 2 of this chapter.

(b) Type acceptance for an individual transmitter may also be requested by an applicant for a station authorization by following the type acceptance procedures set forth in Part 2 of this chapter. Such transmitters, if accepted, will not normally be included on the Commission's "Radio Equipment List", but will be individually enumerated on the station authorization.

(c) Additional rules with respect to type acceptance are set forth in Part 2 of this chapter. These rules include information with respect to withdrawal of type acceptance, modification of type-accepted equipment, and limitations on the findings upon which type acceptance is based.

(d) Transmitters equipped with a frequency or frequencies not listed in § 95.41(d)(1) will not be type accepted for use at Class D stations unless the transmitter is also type accepted for use in the service in which the frequency is authorized, if type acceptance in that service is required.

§ 95.58 Additional requirements for type acceptance.

(a) All transmitters shall be crystal controlled.

(b) Except for transmitters type accepted for use at Class A stations, transmitters shall not include any provisions for increasing power to levels in excess of the pertinent limits specified in Section 95.43.

(c) In addition to all other applicable technical requirements set forth in this part, transmitters for which type acceptance is requested after May 24, 1974, for use at Class D stations shall comply with the following:

(1) Single sideband transmitters and other transmitters employing reduced, suppressed or controlled carrier shall include a means for automatically preventing the transmitter power from exceeding either the maximum permissible peak envelope power or the rated peak envelope power of the transmitter, whichever is lower.

(2) Multi-frequency transmitters shall be capable of operation only on those frequencies authorized by 95.41

(3) All transmitter frequency determining circuitry (including crystals), other than the frequency selection mechanism, employed in Class D station equipment shall be internal to the equipment and shall not be accessible from the exterior of the equipment cabinet or operating panel. Add-on devices, whether internal or external to the equipment, the function of which is to extend the frequency coverage capability of a Class D unit beyond its original frequency coverage capability, shall not be sold, manufactured, or attached to any transmitter capable of operation on Class D Citizens Radio Service frequencies.

(4) Single sideband transmitters shall be capable of transmitting on the upper sideband. Capability for transmission also on the lower sideband is permissible.

(5) The total dissipation ratings, established by the manufacturer of the electron tubes or semiconductors which supply radio frequency power to the antenna terminals of the transmitter, shall not exceed 10 watts. For electron tubes, the rating shall be the Intermittent Commercial and Amateur Service (ICAS plate dissipation value if established. For semiconductors, the rating shall be the collector or device dissipation value, whichever is greater, which may be temperature de-rated to not more than 50°C.

(d) Only the following external transmitter controls, connections or devices will normally be permitted in transmitters for which type acceptance is requested after May 24, 1974, for use at Class D stations. Approval of additional controls, connections or devices may be given after consideration of the function to be performed by such additions.

(1) Primary power connection. (Circuitry or devices such as rectifiers, transformers, or inverters which provide the nominal rated transmitter primary supply voltage may be used without voiding the transmitter type acceptance.)

(2) Microphone connection.

(3) Radio frequency output power connection.

(4) Audio frequency power amplifier output connector and selector switch.

(5) On-off switch for primary power to transmitter. May be combined with receiver controls such as the receiver on-off switch and volume control.

(6) Upper-lower sideband selector; for single sideband transmitters only.

(7) Selector for choice of carrier level; for single sideband transmitters only. May be combined with sideband selector.

(8) Transmitting frequency selector switch.

(9) Transmit-receive switch.

(10) Meter(s) and selector switch for monitoring transmitter performance.

(11) Pilot lamp or meter to indicate the presence of radio frequency output power or that transmitter control circuits are activated to transmit.

(e) An instruction book for the user shall be furnished with each transmitter sold and one copy (a draft or preliminary copy is acceptable providing a final copy is furnished when completed) shall be forwarded to the Commission with each request for type acceptance or type approval. The book shall contain all information necessary for the proper installation and operation of the transmitter including:

(1) Instructions concerning all controls, adjustments and switches which may be operated or adjusted without causing violation of technical regulations of this part;

(2) Warnings concerning any adjustment which, according to the rules of this part, may be made only by, or under the immediate supervision of, a person holding a commercial first or second class radio operator license;

(3) Warnings concerning the replacement or substitution of crystals, tubes or other components which could cause violation of the technical regulations of this part and of the type acceptance or type approval requirements of Part 2 of this chapter.

(4) Warnings concerning licensing requirements and details concerning the application procedures for licensing.

(f) A Class D Citizens Radio Service application form (FCC Form 505), a Temporary Permit, Class D Citizens

Radio Station (FCC Form 555-B), and a copy of Part 95 of the Commission's Rules and Regulations, each to be current at the time of packing of the transmitter, shall be furnished with each transmitter sold after January 1, 1977.

(g) The serial number of each new Class D unit sold after January 1, 1977 shall be engraved on the unit's chassis.

[FR Doc. 76-22424 Filed 8-3-76; 8:45 am]

§ 95.59 Submission of noncrystal controlled Class C station transmitters for type approval.

Type approval of noncrystal controlled transmitters for use at Class C stations in this service may be requested in accordance with the procedure specified in Part 2 of this chapter.

§ 95.61 Type approval of receiver-transmitter combinations.

Type approval will not be issued for transmitting equipment for operation under this part when such equipment is enclosed in the same cabinet, is constructed on the same chassis in whole or in part, or is identified with a common type or model number with a radio receiver, unless such receiver has been certificated to the Commission as complying with the requirements of Part 15 of this chapter.

§ 95.63 Minimum equipment specifications.

Transmitters submitted for type approval in this service shall be capable of meeting the technical specifications contained in this part, and in addition, shall comply with the following:

(a) Any basic instructions concerning the proper adjustment, use, or operation of the equipment that may be necessary shall be attached to the equipment in a suitable manner and in such positions as to be easily read by the operator.

(b) A durable nameplate shall be mounted on each transmitter showing the name of the manufacturer, the type or model designation, and providing suitable space for permanently displaying the transmitter serial number, FCC type approval number, and the class of station for which approved.

(c) The transmitter shall be designed, constructed, and adjusted by the manufacturer to operate on a frequency or frequencies available to the class of station for which type approval is sought. In designing the equipment, every reasonable precaution shall be taken to protect the user from high voltage shock and radio frequency burns. Connections to batteries (if used) shall be made in such a manner as to permit replacement by the user without causing improper operation of the transmitter. Generally accepted modern engineering principles shall be utilized in the generation of radio frequency currents so as to guard against unnecessary interference to other services. In cases of harmful interference arising from the design, construction, or operation of the equipment, the Commission may require appropriate technical changes in equipment to alleviate interference.

(d) Controls which may effect changes in the carrier frequency of the transmitter shall not be accessible from the exterior of any unit unless such accessibility is specifically approved by the Commission.

§ 95.65 Test procedure.

Type approval tests to determine whether radio equipment meets the technical specifications contained in this part will be conducted under the following conditions:

(a) Gradual ambient temperature variations from 0° to 125° F.

(b) Relative ambient humidity from 20 to 95 percent. This test will normally consist of subjecting the equipment for at least three consecutive periods of 24 hours each, to a relative ambient humidity of 20, 60, and 95 percent, respectively, at a temperature of approximately 80° F.

(c) Movement of transmitter or objects in the immediate vicinity thereof.

(d) Power supply voltage variations normally to be encountered under actual operating conditions.

(e) Additional tests as may be prescribed, if considered necessary or desirable.

§ 95.67 Certificate of type approval.

A certificate or notice of type approval, when issued to the manufacturer of equipment intended to be used or operated in the Citizens Radio Service, constitutes a recognition that on the basis of the test made, the particular type of equipment appears to have the capability of functioning in accordance with the technical specifications and regulations contained in this part: *Provided*, That all such additional equipment of the same type is properly constructed, maintained, and operated: *And provided further*, That no change whatsoever is made in the design or construction of such equipment except upon specific approval by the Commission.

95.81 Permissible Communications.

Stations licensed in the citizens radio service are authorized to transmit the following types of communications:

(a) Communications to facilitate the personal or business activity of the licensee.

(b) Communications relating to:
1(1) The immediate safety of life or the immediate protection of property in accordance with 95.85.

(2) The lending of assistance to a motorist, mariner, or other traveler.

(3) Civil defense activities in accordance with 95.12.

(4) Other activities only as specifically authorized pursuant to 95.87.

(c) Communications with other stations authorized in other radio services except as prohibited by 95.83 (a) (3).

95.83 Prohibited Communications.

(a) A citizens radio station shall not be used:

(1) For any purpose or in connection with any activity which is contrary to federal, state, or local law.

(2) For the transmission of communications containing obscene, indecent, or profane words, language, or meaning.

(3) To communicate with an amateur station, an unlicensed station, or foreign stations (other than provided in Subpart E) except for communications pursuant to 95.85 (b) and 95.121.

(4) To convey program material for retransmission live or delayed on a broadcast facility.

> NOTE: A Class A or Class D station may be used in conjunction with administrative, engineering, or maintenance activities of a broadcast station; a Class A or Class C station may be used for control functions by radio which do not involve the transmission of program material; and a Class A or Class D station may be used in the gathering of news items for preparation of news items; *provided* that the actual or recorded transmissions of a citizens radio station are not broadcast at any time in whole or in part.

(5) To intentionally interfere with the communications of another station.

(6) For the direct transmission of any material to the public through a public address system or similar means.

(7) For the transmission of music, whistling, sound effects, or any material for amusement or entertainment purposes or solely to attract attention.

(8) To transmit the word "MAYDAY" or other international distress signals, except when the station is located in a ship, aircraft, or other vehicle which is threatened by grave and imminent danger and requests immediate assistance.

(9) For advertising or soliciting the sale of any goods or services.

(10) For transmitting messages in other than plain language. Abbreviations, including nationally or internationally recognized operating signals, may be used only if a list of all such abbreviations and their meaning is kept in the station records and made available to any commission representative on demand.

(11) To carry on communications for hire whether the remuneration or benefit received is direct or indirect.

(b) A Class D station may not be used to communicate with, or attempt to communicate with, any unit of the same or another station over a distance of more than 150 miles.

(c) A licensee of a Citizens radio station who is engaged in the business of selling Citizens radio transmitting equipment shall not allow a customer to operate under his station license. In addition, all communications by the licensee for the purpose of demonstrating such equipment shall consist only of brief messages addressed to other units of the same station.

§ 95.85 **Emergency and assistance to motorist use.**

(a) All Citizens radio stations shall give priority to the emergency communications of other stations which involve the immediate safety of life of individuals or the immediate protection of property.

(b) Any station in this service may be utilized during an emergency involving the immediate safety of life of individuals or the immediate protection of property for the transmission of emergency communications. It may also be used to transmit communications necessary to render assistance to a motorist.

(1) When used for transmission of emergency communications, certain provisions of this part concerning the use of frequencies (§ 95.41 (d)); prohibited uses (§ 95.83 (a) (3)); operation by or on behalf of persons other than the licensee (§ 95.87); and duration of transmissions (§ 95.81 (a) and (b)) shall not apply.

(2) When used for transmission of communications necessary to render assistance to a traveler, the provisions of this part concerning duration of transmissions (§ 95.91 (b)) shall not apply.

(3) The exemptions granted from certain rule provisions in subparagraphs (1) and (2) of this paragraph may be rescinded by the Commission at its discretion.

(c) If the emergency use under paragraph (b) of this section extends over a period of 12 hours or more, notice shall be sent to the Commission in Washington, D.C., as soon as it is evident that the emergency has or will exceed 12 hours. The notice should include the identity of the stations participating, the nature of the emergency, and the use made of the stations. A single notice covering all participating stations may be submitted.

§ 95.87 Operation by, or on behalf of, persons other than the licensee.

(a) Transmitters authorized in this service must be under the control of the licensee at all times. A licensee shall not transfer, assign, or dispose of, in any manner, directly or indirectly, the operating authority under his station license, and shall be responsible for the proper operation of all units of the station.

(b) Citizens radio stations may be operated only by the following persons, except as provided in paragraph (c) of this section:

(1) The licensee;

(2) Members of the licensee's immediate family living in the same household;

(3) The partners, if the licensee is a partnership, provided the communications relate to the business of the partnership;

(4) The members, if the licensee is an unincorporated association, provided the communications relate to the business of the association;

(5) Employees of the licensee only while acting within the scope of their employment;

(6) Any person under the control or supervision of the licensee when the station is used solely for the control of remote objects or devices, other than devices used only as a means of attracting attention; and

(7) Other persons, upon specific prior approval of the Commission shown on or attached to the station license, under the following circumstances:

(i) Licensee is a corporation and proposes to provide private radiocommunication facilities for the transmission of messages or signals by or on behalf of its parent corporation, another subsidiary of the parent corporation, or its own subsidiary. Any remuneration or compensation received by the licensee for the use of the radiocommunication facilities shall be governed by a contract entered into by the parties concerned and the total of the compensation shall not exceed the cost of providing the facilities. Records which show the cost of service and its nonprofit or cost-sharing basis shall be maintained by the licensee.

(ii) Licensee proposes the shared or cooperative use of a Class A station with one or more other licensees in this service for the purpose of communicating on a regular basis with units of their respective Class A stations, or with units of other Class A stations if the communications transmitted are otherwise permissible. The use of these private radiocommunication facilities shall be conducted pursuant to a written contract which shall provide that contributions to capital and operating expense shall be made on a nonprofit, cost-sharing basis, the cost to be divided on an equitable basis among all parties to the agreement. Records which show the cost of service and its nonprofit, cost-sharing basis shall be maintained by the licensee. In any case, however, licensee must show a separate and independent need for the particular units proposed to be shared to fulfill his own communications requirements.

(iii) Other cases where there is a need for other persons to operate a unit of licensee's radio station. Requests for authority may be made either at the time of the filing of the application for station license or thereafter by letter. In either case, the licensee must show the nature of the proposed use and that it relates to an activity of the licensee, how he proposes to maintain control over the transmitters at all times, and why it is not appropriate for such other person to obtain a station license in his own name. The authority, if granted, may be specific with respect to the names of the persons who are permitted to operate, or may authorize operation by unnamed persons for specific purposes. This authority may be revoked by the Commission, in its discretion, at any time.

(c) An individual who was formerly a citizens radio station licensee shall not be permitted to operate any citizens radio station of the same class licensed to another person until such time as he again has been issued a valid radio station license of that class, when his license has been:

(1) Revoked by the Commission.

(2) Surrendered for cancellation after the institution of revocation proceedings by the Commission.

(3) Surrendered for cancellation after a notice of apparent liability to forfeiture has been served by the Commission.

§ 95.89 Telephone answering services.

(a) Notwithstanding the provisions of § 95.87, a licensee may install a transmitting unit of his station on the premises of a telephone answering service. The same unit may not be operated under the authorization of more than one licensee. In all cases, the licensee must enter into a written agreement with the answering service. This agreement must be kept with the licensee's station records and must provide, as a minimum, that:

(1) The licensee will have control over the operation of the radio unit at all times;

(2) The licensee will have full and unrestricted access to the transmitter to enable him to carry out his responsibilities under his license;

(3) Both parties understand that the licensee is fully responsible for the proper operation of the citizens radio station; and

(4) The unit so furnished shall be used only for the transmission of communications to other units belonging to the licensee's station.

(b) A citizens radio station licensed to a telephone answering service shall not be used to relay messages or transmit signals to its customers.

§ 95.91 Duration of transmissions.

(a) All communications or signals, regardless of their nature, shall be restricted to the minimum practicable transmission time. The radiation of energy shall be limited to transmissions modulated or keyed for actual permissible communications, tests, or control signals. Continuous or uninterrupted transmissions from a single station or between a number of communicating stations is prohibited, except for communications involving the immediate safety of life or property.

(b) All communications between Class D stations (interstation) shall be restricted to not longer than 5 continuous minutes. At the conclusion of this 5-minute period, or the exchange of less than 5 minutes, the participating station shall remain silent for at least 1 minute.

(c) All communication between units of the same Class D station (intrastation) shall be restricted to the minimum practicable transmission.

(d) The transmission of permissible control signals shall be limited to the minimum practicable time necessary to accomplish the desired control or actuation of remote objects or devices. The continuous radiation of energy for periods exceeding 3 minutes duration for the purpose of transmission of control signals shall be limited to control functions requiring at least one or more changes during each minute of such transmission. However, while it is actually being used to control model aircraft in flight by means of interrupted tone modulation of its carrier, a citizens radio station may transmit a continuous carrier without being simultaneously modulated if the presence or absence of the carrier also performs a control function. An exception to the limitations contained in this paragraph may be authorized upon a satisfactory showing that a continuous control signal is required to perform a control function which is necessary to insure the safety of life or property.

§ 95.93 Tests and adjustments.

All tests or adjustments of citizens radio transmitting equipment involving an external connection to the radio frequency output circuit shall be made using a nonradiating dummy antenna. However, a brief test signal, either with or without modulation, as appropriate, may be transmitted when it is necessary to adjust a transmitter to an antenna for a new station installation or for an existing installation involving a change of antenna or change of transmitters, or when necessary for the detection, measurement, and suppression of harmonic or other spurious radiation. Test transmissions using a radiating antenna shall not exceed a total of 1 minute during any 5-minute period, shall not interfere with communications already in progress on the operating frequency, and shall be properly identified as required by § 95.95, but may otherwise be unmodulated as appropriate.

§ 95.95 Station identification.

(a) The call sign of a citizens radio station shall consist of three letters followed by four digits.

(b) Each transmission of the station call sign shall be made in the English language by each unit, shall be complete, and each letter and digit shall be separately and distinctly transmitted. Only standard phonetic alphabets, nationally or internationally recognized, may be used in lieu of pronunciation of letters for voice transmission of call signs. A unit designator or special identification may be used in addition to the station call sign but not as a substitute therefor.

(c) **Except as provided in paragraph (d) of this section, all transmissions from each unit of a citizens radio station shall be identified by the transmission of its assigned call at the beginning and end of each transmission or series of tramsmissions, but at least at intervals not to exceed 10 minutes.**

(d) Unless specifically required by the station authorization, the transmissions of a citizens radio station need not be identified when the station (1) is a Class A station which automatically retransmits the information received by radio from another station which is properly identified or (2) is not being used for telephony emission.

(e) In lieu of complying with the requirements of paragraph (c) of this section, Class A base stations, fixed stations, and mobile units when communicating with base stations may identify as follows:

(1) Base stations and fixed stations of a Class A radio system shall transmit their call signs at the end of each transmission or exchange of transmissions, or once each 15-minute period of a continuous exchange of communications.

(2) A mobile unit of a Class A station communicating with a base station of a Class A radio system on the same frequency shall transmit once during each exchange of transmissions any unit identifier which is on file in the station records of such base station.

(3) A mobile unit of Class A stations communicating with a base station of a Class A radio system on a different frequency shall transmit its call sign at the end of each transmission or exchange of transmissions, or once each 15-minute period of a continuous exchange of communications.

§ 95.97 **Operator license requirements.**

(a) No operator license is required for the operation of a citizens radio station except that stations manually transmitting Morse Code shall be operated by the holders of a third or higher class radiotelegraph operator license.

(b) Except as provided in paragraph (c) of this section, all transmitter adjustments or tests while radiating energy during or coincident with the construction, installation, servicing, or maintenance of a radio station in this service, which may affect the proper operation of such stations, shall be made by or under the immediate supervision and responsibility of a person holding a first- or second-class commercial radio operator license, either radiotelephone or radio telegraph, as may be appropriate for the type of emission employed, and such person shall be responsible for the proper functioning of the station equipment at the conclusion of such adjustments or tests. Further, in any case where a transmitter adjustment which may affect the proper operation of the transmitter has been made while not radiating energy by a person not the holder of the required commercial radio operator license or not under the supervision of such licensed operator, other than the factory assembling or repair of equipment, the transmitter shall be checked for compliance with the technical requirements of the rules by a commercial radio operator of the proper grade before it is placed on the air.

(c) Except as provided in § 95.53 and in paragraph (d) of this section, no commercial radio operator license is required to be held by the person performing transmitter adjustments or tests during or coincident with the construction, installation, servicing, or maintenance of Class C transmitters, or Class D transmitters used at stations authorized prior to May 24, 1974: *Provided*, That there is compliance with all of the following conditions:

(1) The transmitting equipment shall be crystal-controlled with a crystal capable of maintaining the station frequency within the prescribed tolerance;

(2) The transmitting equipment either shall have been factory assembled or shall have been provided in kit form by a manufacturer who provided all components together with full and detailed instructions for their assembly by nonfactory personnel;

(3) The frequency determining elements of the transmitter, including the crystal(s) and all other components of the crystal oscillator circuit, shall have been preassembled by the manufacturer, pretuned to a specific available frequency, and sealed by the manufacturer so that replacement of any component or any adjustment which might cause off-frequency operation cannot be made without breaking such seal and thereby voiding the certification of the manufacturer required by this paragraph;

(4) The transmitting equipment shall have been so designed that none of the transmitter adjustments or tests normally performed during or coincident with the installation, servicing, or maintenance of the station, or during the normal rendition of the service of the station, or during the final assembly of kits or partially preassembled units, may reasonably be expected to result in off-frequency operation, excessive input power, overmodulation, or excessive harmonics or other spurious emissions; and

(5) The manufacturer of the transmitting equipment or of the kit from which the transmitting equipment is assembled shall have certified in writing to the purchaser of the equipment (and to the Commission upon request) that the equipment has been designed, manufactured, and furnished in accordance with the specifications contained in the foregoing subparagraphs of this paragraph. The manufacturer's certification concerning design and construction features of Class C or Class D station transmitting equipment, as required if the provisions of this paragraph are invoked, may be specific as to a particular unit of transmitting equipment or general as to a group or model of such equipment, and may be in any form adequate to assure the purchaser of the equipment or the Commission that the conditions described in this paragraph have been fulfilled.

(d) Any tests and adjustments necessary to correct any deviation of a transmitter of any Class of station in this service from the technical requirements of the rules in this part shall be made by, or under the immediate supervision of, a person holding a first- or second-class commercial operator license, either radiotelephone or radiotelegraph, as may be appropriate for the type of emission employed.

§ 95.101 Posting station license and transmitter identification cards or plates.

(a) The current authorization, or a clearly legible photocopy thereof, for each station (including units of a Class C or Class D station) operated at a fixed location shall be posted at a conspicuous place at the principal fixed location from which such station is controlled, and a photocopy of such authorization shall also be posted at all other fixed locations from which the station is controlled. If a photocopy of the authorization is posted at the principal control point, the location of the original shall be stated on that photocopy. In addition, an executed Transmitter Identification Card (FCC Form 452–C) or a plate of metal or other durable substance, legibly indicating the call sign and the licensee's name and address, shall be affixed, readily visible for inspection, to each transmitter operated at a fixed location when such transmitter is not in view of, or is not readily accessible to, the operator of at least one of the locations at which the station authorization or a photocopy thereof is required to be posted.

(b) The current authorization for each station operated as a mobile station shall be retained as a permanent part of the station records, but need not be posted. In addition, an executed Transmitter Identification Card (FCC Form 452-C) or a plate of metal or other durable substance, legibly indicating the call sign and the licensee's name and address, shall be affixed, readily visible for inspection, to each of such transmitters: *Provided*, That, if the transmitter is not in view of the location from which it is controlled, or is not readily accessible for inspection, then such card or plate shall be affixed to the control equipment at the transmitter operating position or posted adjacent thereto.

§ 95.103 Inspection of stations and station records.

All stations and records of stations in the Citizens Radio Service shall be made available for inspection upon the request of an authorized representative of the Commission made to the licensee or to his representative (see § 1.6 of this chapter). Unless otherwise stated in this part, all required station records shall be maintained for a period of at least 1 year.

§ 95.105 Current copy of rules required.

Each licensee in this service shall maintain as a part of his station records a current copy of Part 95, Citizens Radio Service, of this chapter.

§ 95.107 Inspection and maintenance of tower marking and lighting, and associated control equipment.

The licensee of any radio station which has an antenna structure required to be painted and illuminated pursuant to the provisions of section 303(q) of the Communications Act of 1934, as amended, and Part 17 of this chapter, shall perform the inspection and maintain the tower marking and lighting, and associated control equipment, in accordance with the requirements set forth in Part 17 of this chapter.

§ 95.111 Recording of tower light inspections.

When a station in this service has an antenna structure which is required to be illuminated, appropriate entries shall be made in the station records in conformity with the requirements set forth in Part 17 of this chapter.

§ 95.113 Answers to notices of violations.

(a) Any licensee who appears to have violated any provision of the Communications Act or any provision of this chapter shall be served with a written notice calling the facts to his attention and requesting a statement concerning the matter. FCC Form 793 may be used for this purpose.

(b) Within 10 days from receipt of notice or such other period as may be specified, the licensee shall send a written answer, in duplicate, direct to the office of the Commission originating the notice. If an answer cannot be sent nor an acknowledgment made within such period by reason of illness or other unavoidable circumstances, acknowledgment and answer shall be made at the earliest practicable date with a satisfactory explanation of the delay.

(c) The answer to each notice shall be complete in itself and shall not be abbreviated by reference to other communications or answers to other notices. In every instance the answer shall contain a statement of the action taken to correct the condition or omission complained of and to preclude its recurrence. If the notice relates to violations that may be due to the physical or electrical characteristics of transmitting apparatus, the licensee must comply with the provisions of § 95.53, and the answer to the notice shall state fully what steps, if any, have been taken to prevent future violations, and, if any new apparatus is to be installed, the date such apparatus was ordered, the name of the manufacturer, and the promised date of delivery. If the installation of such apparatus requires a construction permit, the file number of the application shall be given, or if a file number has not been assigned by the Commission, such identification shall be given as will permit ready identification of the application. If the notice of violation relates to lack of attention to or improper operation of the transmitter, the name and license number of the operator in charge, if any, shall also be given.

§ 95.115 False signals.

No person shall transmit false or deceptive communications by radio or identify the station he is operating by means of a call sign which has not been assigned to that station.

§ 95.117 Station location.

(a) The specific location of each Class A base station and each Class A fixed station and the specific area of operation of each Class A mobile station shall be indicated in the application for license. An authorization may be granted for the operation of a Class A base station or fixed station in this service at unspecified temporary fixed locations within a specified general area of operation. However, when any unit or units of a base station or fixed station authorized to be operated at temporary locations actually remains or is intended to remain at the same location for a period of over a year, application for separate authorization specifying the fixed location shall be made as soon as possible but not later than 30 days after the expiration of the 1-year period.

(b) A Class A mobile station authorized in this service may be used or operated anywhere in the United States subject to the provisions of paragraph (d) of this section: *Provided*, That when the area of operation is changed for a period exceeding 7 days, the following procedure shall be observed:

(1) When the change of area of operation occurs inside the same Radio District, the Engineer in Charge of the Radio District involved and the Commission's office, Washington, D.C., 20554, shall be notified.

(2) When the station is moved from one Radio District to another, the Engineers in Charge of the two Radio Districts involved and the Commission's office, Washington, D.C., 20554, shall be notified.

(c) A Class C or Class D mobile station may be used or operated anywhere in the United States subject to the provisions of paragraph (d) of this section.

(d) A mobile station authorized in this service may be used or operated on any vessel, aircraft, or vehicle of the United States: *Provided*, That when such vessel, aircraft, or vehicle is outside the territorial limits of the United States, the station, its operation, and its operator shall be subject to the governing provisions of any treaty concerning telecommunications to which the United States is a party, and when within the territorial limits of any foreign country, the station shall be subject also to such laws and regulations of that country as may be applicable.

§ 95.119 Control points, dispatch points, and remote control.

(a) A control point is an operating position which is under the control and supervision of the licensee, at which a person immediately responsible for the proper operation of the transmitter is stationed, and at which adequate means are available to aurally monitor all transmissions and to render the transmitter inoperative. Each Class A base or fixed station shall be provided with a control point, the location of which will be specified in the license. The location of the control point must be the same as the transmitting equipment unless the application includes a request for a different location. Exception to the requirement for a control point may be made by the Commission upon specific request and justification therefor in the case of certain unattended Class A stations employing special emissions pursuant to § 95.47(e). Authority for such exception must be shown on the license.

(b) A dispatch point is any position from which messages may be transmitted under the supervision of the person at a control point who is responsible for the proper operation of the transmitter. No authorization is required to install dispatch points.

(c) Remote control of a Citizens radio station means the control of the transmitting equipment of that station from any place other than the location of the transmitting equipment, except that direct mechanical control or direct electrical control by wired connections of transmitting equipment from some other point on the same premises, craft, or vehicle shall not be considered remote control. A Class A base or fixed station may be authorized to be used or operated by remote control from another fixed location or from mobile units: *Provided*, That adequate means are available to enable the person using or operating the station to render the transmitting equipment inoperative from each remote control position should improper operation occur.

(d) **Operation of any Class C or Class D station by remote control is prohibited except remote control by wire upon specific authorization by the Commission when satisfactory need is shown.**

§ 95.121   **Civil defense communications.**

A licensee of a station authorized under this part may use the licensed radio facilities for the transmission of messages relating to civil defense activities in connection with official tests or drills conducted by, or actual emergencies proclaimed by, the civil defense agency having jurisdiction over the area in which the station is located: *Provided*, That:

(a) The operation of the radio station shall be on a voluntary basis.

(b) [Reserved]

(c) Such communications are conducted under the direction of civil defense authorities.

(d) As soon as possible after the beginning of such use, the licensee shall send notice to the Commission in Washington, D.C., and to the Engineer in Charge of the Radio District in which the station is located, stating the nature of the communications being transmitted and the duration of the special use of the station. In addition, the Engineer in Charge shall be notified as soon as possible of any change in the nature of or termination of such use.

(e) In the event such use is to be a series of pre-planned tests or drills of the same or similar nature which are scheduled in advance for specific times or at certain intervals of time, the licensee may send a single notice to the Commission in Washington, D.C., and to the Engineer in Charge of the Radio District in which the station is located, stating the nature of the communications to be transmitted, the duration of each such test, and the times scheduled for such use. Notice shall likewise be given in the event of any change in the nature of or termination of any such series of tests.

(f) The Commission may, at any time, order the discontinuance of such special use of the authorized facilities.

SUBPART E—OPERATION OF CITIZENS RADIO STATIONS IN THE UNITED STATES BY CANADIANS

§ 95.131 Basis, purpose and scope.

(a) The rules in this subpart are based on, and are applicable solely to the agreement (TIAS #6931) between the United States and Canada, effective July 24, 1970, which permits Canadian stations in the General Radio Service to be operated in the United States.

(b) The purpose of this subpart is to implement the agreement (TIAS #6931) between the United States and Canada by prescribing rules under which a Canadian licensee in the General Radio Service may operate his station in the United States.

§ 95.133 Permit required.

Each Canadian licensee in the General Radio Service desiring to operate his radio station in the United

States, under the provisions of the agreement (TIAS #6931), must obtain a permit for such operation from the Federal Communications Commission. A permit for such operation shall be issued only to a person holding a valid license in the General Radio Service issued by the appropriate Canadian governmental authority.

§ 95.135 Application for permit.

(a) Application for a permit shall be made on FCC Form 410–B. Form 410–B may be obtained from the Commission's Washington, D.C., office or from any of the Commission's field offices. A separate application form shall be filed for each station or transmitter desired to be operated in the United States.

(b) The application form shall be completed in full in English and signed by the applicant. The application must be filed by mail or in person with the Federal Communications Commission, Gettysburg, Pa. 17325, U.S.A. To allow sufficient time for processing, the application should be filed at least 60 days before the date on which the applicant desires to commence operation.

(c) The Commission, at its discretion, may require the Canadian licensee to give evidence of his knowledge of the Commission's applicable rules and regulations. Also the Commission may require the applicant to furnish any additional information it deems necessary.

§ 95.137 Issuance of permit.

(a) The Commission may issue a permit under such conditions, restrictions and terms as it deems appropriate.

(b) Normally, a permit will be issued to expire 1 year after issuance but in no event after the expiration of the license issued to the Canadian licensee by his government.

(c) If a change in any of the terms of a permit is desired, an application for modification of the permit is required. If operation beyond the expiration date of a permit is desired an application for renewal of the permit is required. Application for modification or for renewal of a permit shall be filed on FCC Form 410-B.

(d) The Commission, in its discretion, may deny any application for a permit under this subpart. If an application is denied, the applicant will be notified by letter. The applicant may, within 30 days of the mailing of such letter, request the Commission to reconsider its action.

§ 95.139 **Modification or cancellation of permit.**

At any time the Commission may, in its discretion, modify or cancel any permit issued under this subpart. In this event, the permittee will be notified of the Commission's action by letter mailed to his mailing address in the United States and the permittee shall comply immediately. A permittee may, within 30 days of the mailing of such letter, request the Commission to reconsider its action. The filing of a request for reconsideration shall not stay the effectiveness of that action, but the Commission may stay its action on its own motion.

§ 95.141 **Possession of permit.**

The current permit issued by the Commission, or a photocopy thereof, must be in the possession of the operator or attached to the transmitter. The license issued to the Canadian licensee by his government must also be in his possession while he is in the United States.

§ 95.143 **Knowledge of rules required.**

Each Canadian permittee, operating under this subpart, shall have read and understood this Part 95, Citizens Radio Service.

§ 95.145 **Operating conditions.**

(a) The Canadian licensee may not under any circumstances begin operation until he has received a permit issued by the Commission.

(b) Operation of station by a Canadian licensee under a permit issued by the Commission must comply with all of the following:

(1) The provisions of this subpart and of Subparts A through D of this part.

(2) Any further conditions specified on the permit issued by the Commission.

§ 95.147 Station identification.

The Canadian licensee authorized to operate his radio station in the United States under the provisions of this subpart shall identify his station by the call sign issued by the appropriate authority of the government of Canada followed by the station's geographical location in the United States as nearly as possible by city and state.

Index

A

AC wiring, base station	154
Address change	34
Age, license	34
Angle of radiation	97, 100
Antennas	
checks	144, 158
directory	111
frequency sensitivity	102
height	109
rotators	95
Antenna switch	154
Anti-static powder	142
Applications, CB radio	15
license	34
Attenuation, coax cable	127
Auto antenna	94
Auto electrical system	136

B

Base station	10, 48
Base station antennas	97
installation	154
Battery check	158
Battery connections, mobile	134
Battery operated transceiver	46
Battery polarity	136
Beam antenna	94
Bumper mount	136
Business CB uses	15

C

Capacity, coax cable	130
Carrier	158
Centerloaded antennas	95
Change of address	34
Citizens radio service	203
Class D license	11
Cleaning	166
Coast Guard channels	181
Coax cable	126
connectors	134
selection	133
Coaxial antennas	95
Coaxial switch	153
Coil-loaded antennas	95
Coil, noise	152
Color, cable dielectric	128
Community radio watch	32
Compressor, speech	163
Connectors, coax cable	134
Construction CB uses	18
Contact cleaning	166
Continuity check, mobile antenna	151

D

DAIR	28
Dead receiver	172
Dead transceiver	171
Dead transmitter	174
Dielectric constants	128

Dielectric, coax cable	128
Dipole antenna	100
Directional antennas	97
Directivity, antenna	107
Directors, antenna	107
Distortion	173
Distributor, noise	131
Double-conversion	48
Drilling mounting holes	136
Driving aid, information, and routing	28
DX	79

E

Effective radiated signal	98
Electrical system, auto	131
Eligibility, license	34
Emergency base station antenna	187
Emergency frequency	179
Emergency monitoring station	178

F

Farm CB uses	16
FCC Form 505	36
FCC mailing address	44
FCC Rules	193
Fee, license	36
Fixed service	10
Foot switch, microphone	158
Form, license	36
Frequency, coax loss related to	126
Frequency sensitivity, antenna	166
Frequency tolerance	11
Front-to-back ratio, antenna	107
Fuse	171

G

Gauges, noise	151
Generator noise	152
Grounding, base station	158
Ground plane antenna	94, 102
Grounds, base station	155
Groundwaves	110
Guying, antenna	154

H

Half-wave dipole	107
Hand-held transceiver	46, 74
HART	27
HELP	23
Highway emergency locating plan	23
Highway emergency radio	27
Home CB uses	16

Horizontally polarized antenna	98
Hum	174

I

Industrial CB uses	15
Installation, base station antenna	158
Installation, mobile	136
Instructions, license applications	37
Intermittent operation	173
Inventory control CB uses	21

J

Jamming	77

L

Land station	10
Law enforcement CB uses	24
Lead-in wire	108
License classes	9
"Licensed" towers	158
License fee	36
License-free transceiver	46, 74
License modification	35
Licensing requirements	34
"Lock-in" transmitting	83
Long-distance communications	80
Losses, coax cable	108, 126
Low-pass filter	162

M

Mailing address, FCC	44
Maintenance, transceivers	166
Marine CB uses	16
Marine distress	181
Microphone	164
Microphone switch	160
Military specifications, coax cable	127
Mismatch, antenna	107
Mobile antenna	94, 143
Mobile antenna continuity check	153
Mobile communications van	185
Mobile equipment	48
Mobile service	10
Model control	11
Modification	35
Modulation	71, 163, 174
Monitoring stations, emergency	178
Monitor marker signal	189
Monitor, power	160
Monitor receiver	181

255

Motors, noise	152
Mounting base station antennas	154
Mounting, mobile antenna	143
Mobile transceiver	136
Multiple-channel operation	22

N

"Night-time factor"	81
Noise suppression	151
Non-interference communications	74
Non-license transceiver	46, 74

O

Omni-directional antenna	100
Opacity, coax cable	130
Operating requirements	11

P

Parasitic elements, antenna	107
Part 15 rules	87
Part 15 tranceivers	46, 74
Part 95 rules	203
Polarization	110
Police CB uses	24
Police/Fire radio directory	184
Power	11
Power monitor	161
Power supply checks	171
Power supply, mobile	164
Preventive maintenance	166
Propagation	110
Public service CB uses	23

R

Radiation angle	97, 100
Radio emergency associated citizens teams	26, 175
Radio waves	110
Range, CB operating	108
REACT	26, 175
Receiver troubles	172
Recreational CB uses	16
Reflected power	108
Remote control	11
Renewal, license	34
Resistance, mobile antenna lead-in	153
RF attenuation	132
"RG" cable	126
Rotator, antenna	95

S

"Second" transceiver	181
Selective-calling	11, 164

Selectivity	51
Sensitivity	50
Service industry CB uses	23
Servicing	167
Sidebands	71, 164
Sign, monitor marker	189
Single-sideband	71
"Skip" communications	10, 80, 110
"Sky waves"	94, 110
Spacing, yagi antenna elements	107
Spark plugs, noise	151
Speech compressor	163
Sporadic-E skip	80
Spring mount, antenna	95
Squelch	48
SSB	71
Stacked yagis	107
Standing waves	107
Superhet receiver	50
Switch cleaning	166

T

"Thunderstick" antenna	113
Tires, noise	152
Tolerance, frequency	11
Tone-signaling	11, 83, 165
Transceiver maintenance	166
Transmission line	126, 133
Troubleshooting	167
Tunable receiver	48
Tuning, whip antenna	106

U

UHF stations	11
Unidirectional antennas	107
Used equipment	67
Uses, CB radio	15

V

Vertically-polarized antennas	100
Violations	45
Voltage regulator, noise	152
VSWR	106, 108

W

Walkie talkies	50, 72
Weak modulation	174
Weak reception	173
Weatherproofing	158
Whip antennas	95, 101
Wiring, base station	160

Y

Yagi antenna	107